Proceedings of the *Académie Internationale de Philosophie des Sciences*

Comptes Rendus de l'Académie Internationale de Philosophie des Sciences

Tome II

Justification, Creativity, and Discoverability in Science

Tome I
Science's Voice of Reflection
Gerhard Heinzmann and Benedikt Löwe, Éditeurs

Tome II
Justification, Creativity, and Discoverability in Science
Lorenzo Magnani, Éditeur

Justification, Creativity, and Discoverability in Science

Éditeur

Lorenzo Magnani

© Individual author and College Publications 2024
All rights reserved.

ISBN 978-1-84890-474-3

College Publications, London
Scientific Director: Dov Gabbay
Managing Director: Jane Spurr

http://www.collegepublications.co.uk

Original cover design by Laraine Welch

All rights reserved. No part of this publication may be reproduced, stored in a retrieval system or transmitted in any form, or by any means, electronic, mechanical, photocopying, recording or otherwise without prior permission, in writing, from the publisher.

Table of Contents

Preface
Lorenzo Magnani   vii–viii

Second-level evidence for future-proof science?
Mario Alai   1–13

Serendipity between psychology and logic of scientific discovery
Marco Buzzoni   15–38

Realism, scientific creativity, and theory justification
Alberto Cordero   39–53

Plausible hypothesis constructed by abduction: some examples in sciences
Jean-Pierre Desclés   55–72

Continuity and discontinuity in theory change
Dennis Dieks   73–84

Justifying scientific beliefs: an anti-naturalist and anti-pragmatist perspective
Michel Ghins   85–102

The concept of creativity in the sciences
Reflections on some problems
Hans-Peter Grosshans   103–113

Justification, creativity and discoverability in mathematics: the example of predicativity
Gerhard Heinzmann   115–123

Discoverability: affordances and eco-cognitive situatedness
Towards an ecology of human creativity
Lorenzo Magnani   125–139

On scientific creativity and its limitations
Fabio Minazzi   141–148

Hypothetical value judgements: Reconciling value-neutrality and
value-engagement in science
Gerhard Schurz   149–178

Preface

This volume is a collection of selected papers that were presented at the international conference of the *Académie Internationale de Philosophie des Sciences* (AIPS) entitled *Justification, Creativity, and Discoverability in Science*, and held in the *Palazzo Vistarino* of the *Università di Pavia* in Italy from 26 September to 1 October 2022, chaired by the editor of these proceedings on behalf of the AIPS (cf. Figure 1).

The presentations delivered at the conference in Pavia delved into the role of justification and creativity within scientific thought, also aiming to instigate the innovative problem of *discoverability* that affects various areas of current scientific enterprise changes in theories and concepts. Some speakers tackled the intricacies of justification and creativity within science, mathematics, and technology, emphasizing the complex relationship between scientific exploration and technological advancement, the ontological regimes, the role of abduction, induction, and thought experiments, theory change, the role of belief and of justification of axioms. The following is the list of presentations given at the conference in alphabetic order; presentations that are represented by a paper in this volume are marked by an asterisk ⋆.

⋆**Alai, Mario**. *Can we recognize future-proof science, and how?*

⋆**Buzzoni, Marco**. *Method, creativity, and serendipity in scientific research.*

⋆**Cordero, Alberto**. *Ontological commitment, creativity, and justification in science.*

⋆**Desclés, Jean-Pierre**. *Plausible hypothesis constructed by abduction: some examples of discovery in sciences.*

⋆**Dieks, Dennis**. *Continuity and discontinuity in theory change.*

Fano, Vincenzo. *Thought experiments in empirical science. Necessary but unreliable.*

⋆**Ghins, Michel**. *Justifying scientific beliefs: an anti-pragmatist and anti-naturalist perspective.*

⋆**Grosshans, Hans-Peter**. *The concept of creativity in respect to the sciences—Reflections on some problems.*

⋆**Heinzmann, Gerhard**. *Justification, creativity and discoverability in mathematics: the example of predicativity.*

Kahle, Reinhard. *Justifying axioms.*

⋆**Magnani, Lorenzo**. *Discoverability—The critical need for and ecology of human creativity.*

⋆**Minazzi, Fabio**. *On scientific creativity and its constraints.*

Rheinberger, Hans-Jörg. *On the moment of creativity in science—Two vistas*

FIGURE 1. Palazzo Bellisomi Vistarino, remodeled by Francesco Croce (1696–1773) in the mid-18th century in the style of Lombard Rococo for the aristocratic Bellisomi family. *Left.* First courtyard. Picture: Fabio Romanoni, October 2020. *Right.* Frescoes by Giovanni Angelo Borroni (1684–1772). Picture: Università di Pavia, 2018.

⋆**Schurz, Gerhard**. *The optimality of meta-induction: A new account to Hume's problem.*
Zovko, Jure. *The role of judgment in scientific discovery.*

The editor of these proceedings expresses his appreciation to Jure Zovko, Ivana Nicolić, Benedikt Löwe, and Thomas Piecha, the President and Secretary of AIPS and the editor and the typesetter of their *Comptes Rendus*, respectively, for their help in the organization of the conference and the preparation of the present volume.

The conference and thus indirectly this volume were made possible through the generous financial support of the MUR (Italian Ministry of the University: PRIN 2017 Research 20173YP4N3), of the *Dipartimento di Studi Umanistici* of the *Università di Pavia*, and of the *Académie Internationale de Philosophie des Sciences*. Their support is gratefully acknowledged. The preparation of the volume would not have been possible without the contribution of resources and facilities of the Computational Philosophy Laboratory and of the *Dipartimento di Studi Umanistici*, Philosophy Section of the *Università di Pavia*.

Sestri Levante
October 2024

L.M.

Second-level evidence for future-proof science?

Mario Alai

Università degli Studi di Urbino Carlo Bo, via Curiel 41, 47521 Cesena, Italy

1 Can we identify securely true hypotheses in current science?

According to the "pessimistic meta-induction" none of our current scientific theories, hypotheses or assumptions are true and will be preserved in the future. For some hyper-optimistic outlooks (e.g., Doppelt 2007, 2011), practically all current science, unlike past science, is true and (save minor adjustments) it will stay forever. The latter view runs against fallibilism, and the former against the hope that through science we can reach at least some truths. However, the current debates on scientific realism abundantly show that both extremes are wrong: most realists are nowadays selective realists, but also many antirealists are only *selectively* so. This means that we now possess at least some truths. Moreover, if science is progressive, or at least not badly regressive, it follows that many, or at least some, of our true scientific claims are future-proof, i.e., they will never be refuted.

The question, is, however: Can we identify which ones? Clearly, not all of those we hold now, for that would mean that now we are infallible. Moreover, it would seem that to distinguish which of our claims are true and enduring, and which ones are false, we should be able to anticipate future scientific progress, which is impossible. Yet, this question is becoming crucial today, not only for philosophers or historians of science but also for policymakers and the general public: the COVID-19 pandemic has shown how important it is for our very safety that even individual laypersons become able to distinguish between mere scientific *opinions* and established scientific *facts*.

Some, like Alberto Cordero (2017a, 2017b), maintain that for certain claims we now have such overwhelming evidence that it would be nonsensical to imagine that they can be rejected in the future. For instance, despite wide disagreements on the interpretations of quantum mechanics, physicists of all leanings and allegiances agree on certain basic tenets (Cordero 2001, p. 307). But who is to say when evidence is enough to warrant that we are in front of an indisputable *fact*? Earlier on, Rescher (1987, Ch. 5) distinguished between "forefront science", which is precise but mostly false, and "schoolbook science", which though vague and imprecise includes the true core of forefront science. But this distinction is vague; moreover, it doesn't seem to capture the distinction between reversible and perduring claims, because some of today's forefront science will still be true in the future, while much science that was in the schoolbooks of the past has subsequently been rejected.

Deployment realists (notably, Psillos 1999) have convincingly used the "no miracle" argument from novel predictions (henceforth 'NMA') to argue that when a hypothesis was essentially deployed in a novel and risky prediction, we can be practically certain of its truth; but from this it is a short step to conclude that (save unfortunate but unlikely scientific regressions) it is future-proof. There is a problem, however: to begin with, we must distinguish between claims that are *completely* true (i.e., true *tout court*) and others that are only *partly* true (i.e., false but with some true content, i.e., consequences). For instance,

(SW) All swans are white

is false, for some Australians swans are black. However, it entails the true statements that all European swans are white, all American swans are white, all Asian swans are white, and all African swans are white. These latter four claims are completely true, while (SW) is only partly true. This is what we mean by saying that (SW) is *approximately* true, and it explains the remarkable empirical and practical success that (SW) provides to its holders (Musgrave 2006–7). If we trust in the progressive nature of science, we can expect a completely true claim to be future-proof, while a partly true claim, in the long run, will (hopefully) be rejected and substituted by its completely true parts (i.e., consequences).

Now, when a hypothesis H is used to derive a novel risky prediction, there are two possible cases: either it has been deployed *essentially*, hence most probably it is completely true, or it was deployed *inessentially*, hence it is only partially true (Alai 2014a, § 7; 2021). H is deployed inessentially in a prediction when the latter was actually deduced from it, but it might equally well have been deduced just from a part of it. In that case, only the essential part of H is certainly true. For instance, (SW) may be employed inessentially in deriving the prediction

(PR) Any swan I see in Urbino will be white,

for the same prediction may also be derived from some of its parts, like 'All European swans are white', or 'All Italian swans are white', etc. Still, even the latter statements would be inessential to that prediction, while *in practice* only something like

(UrSW) All swans in Urbino are white

might count as essential.[1] For instance, as argued by Psillos (1999, 121), we need not be committed to the caloric hypothesis, although it was actually

---

[1] It may be remarked that, from a purely logical point of view, only (PR) itself is essential to deriving (PR). However, by hypothesis the fact described by a prediction (PR)

deployed in Laplace's prediction of the speed of sound in the air because it was not deployed essentially, hence that prediction "did not depend on this hypothesis". The same goes for the existence of the ether and other false assumptions that were deployed in successful novel predictions, albeit inessentially (Alai 2014a, § 7, 2021, §§ 9.4, 9.5). Therefore, only the assumptions that have played an *essential* role in novel scientific predictions can be trusted to be *completely* (not just *partially*) true, hence destined for preservation in the long run.

However, whenever a novel prediction np has been derived from a hypothesis H, it is *practically* impossible to tell whether H was employed essentially or not, and if not, which part of H was essential. One reason is that which is the minimal assumption a scientist needs to predict np depends on her background knowledge. Only in hindsight, if H is found to be false by subsequent research, it is shown that H had not been essential (Alai 2021, § 9.5). It follows that we cannot exactly circumscribe future-proof claims: the most we can learn from the NMA is that a hypothesis is at least partly true, and this doesn't guarantee that it will be preserved in the long run. Of course, this is to say that we do have some completely true (hence future-proof) beliefs (those that were actually essential in deriving the relevant predictions), but we don't know which ones.

Moreover, in his book *Identifying Future-Proof Science* (2022), Peter Vickers maintains that the question whether a claim is actually new and risky seldom, if ever, allows for a clear-cut answer, and it can be decided only through interdisciplinary competencies. In the past, I too argued that novelty, riskiness, and inessentiality are gradual properties (2014b, §§ 3.4, 4). Summing up, in many cases the NMA can be of little help in identifying future-proof claims.

More generally, Vickers points out that the first-level empirical evidence one would need to assess in order to decide whether a claim is future-proof is so vast and requires such specialized competence in different disciplines that not only no philosopher or layperson, but not even an individual scientist could master all of it. Even if a scientist could, over many years, study all that material, he would still see it from his individual and potentially biased perspective.

2 Vickers' criterion for future-proof scientific claims

Although neither the NMA nor the direct assessment of any other first-level empirical evidence allows us to decide whether a claim is future-proof, Vickers argues that this can be decided by a second-level criterion:

---

is unknown to us, and we must derive it from some assumption (H) we know. Moreover, if (PR) is to be deduced from (H), (H) must be stronger than (PR). Yet, there are problems in isolating essential assumptions: see the next paragraph.

(C) If the community of scientists competent concerning a claim C is sufficiently large and representative of different perspectives and sociological groups, and at least 95% of its members believe that C describes an established scientific fact (i.e., beyond reasonable doubt, save philosophical skepticism), then C is future-proof.

Since checking whether a given claim complies with (C) is in principle possible even for philosophers or laypersons, (C) may allow us to recognize future-proof claims even for the needs of philosophy and practical life.[2] Moreover, (C) is only a sufficient condition of truth and permanence, not a necessary one: unbeknown to us, even claims that don't command a 95% consensus, or do not yet, might be future-proof.

Granted, (C) runs against the current wisdom of both laypersons and philosophers: as famously argued by Kuhn (1962), scientific consensus may be achieved even for sociological (non-epistemic) reasons, and many theories or hypotheses that in the past were accepted as a matter of course have subsequently been rejected by the "scientific revolutions" and are now considered straightforwardly false. Yet, Vickers holds this criterion is borne out by the history of science: no claim fulfilling the requirements of (C) has ever been rejected: all the once largely accepted claims that have subsequently been rejected were still debated by at least some 6% of specialized scientists. Moreover, Vickers lists 30 statements or whole bodies of knowledge which met criterion (C) a long time ago, (pp. 12–18, 220).

However, one might grant that (C) has been confirmed so far, but ask whether this is only a contingent fact, or we can be assured that also present and future claims fulfilling the conditions of (C)'s antecedent will be future-proof. In other words, one may ask whether (C) is predictively reliable, and an affirmative answer may be provided only by *justifying* (C): *why* can we trust that claims fulfilling (C)'s antecedent are future-proof?

To begin with, Vickers remarks that in some cases almost unanimous consensus is reached when, thanks to technical progress, the entities, structures, or behaviors that were originally unobservable become directly observable by means of appropriate instruments, as happened with continental drift and the SARS-CoV-2 virus. But, he adds, scientific debates on the reliability of instruments, like the question of whether we see through optical microscopes, were resolved many years ago (pp. 198–200, 221). Therefore, claims that are

---

[2] In practice this may be more difficult: as Vickers remarked in discussion, some scientists are not convinced that global warming is anthropogenic, but they are very few. More or less than 5%? Of course, there aren't exact figures, moreover, it depends on which scientists we count as competent on this matter. Therefore, some have objected that in practice checking whether a claim satisfies (C) can be so hard, that it isn't actually helpful. Yet, while in some cases it may be difficult, in other cases it is certainly easier. Besides, even if it may be difficult to give an *exact* answer, in many cases it will be possible to give an approximate one, and for practical purposes that may be enough.

unanimously accepted on such bases are certainly true and destined to be preserved in the future.

As to the claims about entities that are not even instrumentally observable, it must be considered that science is essentially a critical activity, with strong epistemic and sociological premiums on criticism and nonconformism; therefore, if any doubts about a given claim were still possible, at least a substantial minority of scientists (well over 5%) would have raised them. "Any solid international scientific consensus is so hard-won that the evidence base has to be truly enormous to achieve it" (p. 112). As remarked by Kuhn himself, "History suggests that the road to a firm research consensus is extraordinarily arduous" (1962, p. 15). This is why if at least 95% of scientists have no doubts about a claim, it must be future-proof.

Still, it may be objected that various sociological and epistemic forces in scientific practice push to conformity, and Vickers is ready to grant that these forces may be even stronger today than in Kuhn's times. Therefore, when a claim gains acceptance, the "bandwagon" effect is a possible threat (p. 220). Thus, in the end, Vickers gives up on providing a full principled explanation of *how* scientists may reach a 95% consensus, and *why* it is such a reliable indicator of truth and stability; instead, he settles for just maintaining *that* it is, by induction from its confirmation in the history of science so far.

3 Why is Vickers' criterion reliable?

However, if a 95% consensus could be reached for merely sociological reasons (or for any epistemically irrelevant reason), the fact that claims passing the (C) standard have been preserved *so far* wouldn't be a sufficient reason to assume that they are true and that they will always be accepted in the future, i.e., the inductive support for (C) would become weaker, or even irrelevant. One reason to suspect that even claims that now command a 95% consensus might not be really future-proof is Stanford's (2006) idea that at any time, hence even now, there are alternatives to accepted theories that escape us, one of which might be the true answer to our questions. Another reason is the abovementioned fact that hypotheses which now seem to have been essential in deriving certain novel predictions (hence completely true) may actually have been inessential. Therefore, the questions of how the 95% consensus is reached and why it is reliable remain compelling.

Thus, in order to establish the reliability of criterion (C) one should show

(A) which *epistemically* relevant reasons (i.e., relevant to the truth of a claim) may produce a 95% consensus,

and

(B) why epistemically irrelevant reasons cannot produce such a consensus.

As mentioned earlier, a clear and satisfactory answer to question (A) is provided by Vickers for the claims about facts that have become observable thanks to new sophisticated instruments. Here I shall try to explore some further steps toward answering this question in the general case, and in the case of non-observable facts.

It is usually assumed that the probability conferred by some empirical evidence e to a hypothesis H is given by Bayes' theorem:

(i) $p(H|e) = \dfrac{p(e|H) \cdot p(H)}{p(e)}$

or, in its extended formulation,

(ii) $p(H|e) = \dfrac{p(e|H) \cdot p(e)}{[p(e|H) \cdot p(H)] + [p(e|\neg H) \cdot p(\neg H)]}.$

However, according to the NMA, the success of novel predictions also provides some second-level evidence for H, and this is why: all true hypotheses have only true consequences, while not all the consequences of false hypotheses are true: in fact, most false hypotheses have mostly false consequences. For instance, if it is true that

H Tomorrow is Tuesday,

all the consequences of H are also true:

C_1 Today is Monday.

C_2 Yesterday was Sunday.

C_3 The day after tomorrow is Wednesday.

C_n Etc.

C_{n+1} Today is a weekday.

C_{n+2} Tomorrow's name's initial is T.

C_{n+3} Etc.

Instead, consider the six false hypotheses about which day tomorrow is: none have the true consequences C_1 to C_n; besides, only five of them have C_{n+1}, only one has C_{n+2}, etc.

Now, theoreticians look for true hypotheses, and they try to build them in order to entail certain known evidence e, e', e'', etc. Hence their hypotheses, even if false, will entail e, e', e''. However, it is only by chance that they build a false hypothesis with a true *novel* and unlikely consequence ne. Therefore, we may ask what the probability is that by chance one picks a false hypothesis H with a true consequence ne. Clearly, it is equal to the ratio of false hypotheses that entail ne to all the false hypotheses on that subject.

In turn, that ratio is inversely proportional to the logical content of ne: if ne is tautological all hypotheses entail it, while if it is contradictory none do; between these two extremes, the greater the logical content of ne, the fewer the hypotheses that entail it. We may measure the logical content of ne as the ratio of the number of equiprobable cases it excludes to the number of all equiprobable cases (where equiprobable cases are those which we have no reason to believe have different probabilities).

Conversely, we may call the "logical probability" of ne the ratio of the number of equiprobable cases it allows to the number of equiprobable cases: if ne is tautological (allowing all cases) its logical probability is 1, and if it is contradictory (excluding all cases) it is 0. Hence, the probability that by chance one picks a false hypothesis H entailing ne is equal to the logical probability of ne.

Determining which are the equiprobable cases is not always possible, and it involves a principle of indifference which depends on our background assumptions. However, in several actual scientifically relevant cases one can figure out in a substantially clear way what the possible alternative cases are, hence what the logical probability of ne is.

For instance, when Adams and Leverrier predicted the mass and position of a new planet (later called Neptune), the equiprobable cases were the other positions and masses that a new planet might possibly have had. As for the mass, it is difficult to tell what the possible equiprobable alternatives were (e.g., those not obviously too large or too small). However, their prediction of the position missed the mark by just 1° over 360°, so their true prediction was $n° \pm 1°$, hence its logical probability was 2/360, or 1/180. This is to say that only one in 180 logically possible (groups of) hypotheses entailing predictions of Neptune's position entailed the right one. Therefore, the logical probability that a theoretician picked a hypothesis entailing the right prediction *by chance* was also 1/180.

Now, only by chance a false hypothesis entails a true consequence. Hence, the probability that Adams and Leverrier predicted Neptune's position using false hypotheses (i.e., that the hypotheses of Newton's gravitation theory essentially used by them were false) is 1/180. On the opposite, the probability $p(H/ne)$ that they made this prediction using true hypotheses (i.e., that the hypotheses of Newton's gravitation theory essentially used by them were true) is fairly high. Granted, if the antecedent probability of Newton choosing false hypotheses $p(H)$ were very low, also $p(H/ne)$ would be low. But in general, this is not the case (Alai, 2023).

Since the probability of deriving a novel prediction ne from a false hypothesis H is proportional to ne's logical probability lp(ne), when ne is very risky (i.e., when lp(ne) is very low) it would be a "miracle" if H were false.

Besides, many contemporary novel predictions are much riskier and much more approximate than the prediction of Neptune's position, hence the probability that the hypotheses used to derive them are true is proportionally much larger: for instance, quantum electrodynamics predicted the magnetic moment of the electron to be $1159652359 \times 10^{-12}$, while experiments found $1159652410 \times 10^{-12}$: hence John Wright (2002, pp. 143–144) figured that the probability to get such accuracy *by chance*, i.e., through a false theory, is as low as 5×10^{-8}.

Thus, in the ideal case of NMA (when the logical probability of ne can be assessed with sufficient confidence) we might recognize that H is (at least partly) true from just one piece of evidence (the confirmation of ne). Something similar, perhaps, happened with Eddington's confirmation of Einstein's prediction of the bending of light in the 1921 solar eclipse. Hence, this is a kind of epistemically relevant consideration which may underly, explain and warrant a 95% consensus.

It might be objected that, just because there are infinitely many false hypotheses for any true one, the probability of finding a true hypothesis entailing ne is still lower than that of finding a false one entailing ne. Indeed, this would be the case if all hypotheses were picked by chance. But in that case, even the probability of picking a false hypothesis entailing a true risky prediction would be minimal, definitely too low to explain the many extraordinary successful predictions of our scientific theories. Therefore, we must conclude that scientists do not pick hypotheses randomly, but seek true hypotheses (which necessarily entail true consequences) and sometimes find them, and this happens because they employ reliable heuristics. This is why hypotheses that license novel risky predictions are most probably true (White 2003; Alai 2014c, § 6).

Bayes' theorem establishes the conditional probability of H on the basis of the antecedent probability of H, the antecedent probability of e, and the logical relationship of H to e. In general, the antecedent probabilities of H and e are different from their logical probabilities, for they are based on certain prior empirical or theoretical evidence we have for them. Only when we have no prior information, do their antecedent probabilities reduce to their logical probabilities, but this is seldom or practically never the case. Thus, Bayes' theorem assesses the conditional probability of H with first-level evidence.

Instead, the NMA assesses the probability that H is true only by considering that it has licensed a novel risky prediction ne, irrespective of what H

and ne say and of the first-level (empirical or theoretical) evidence we have for them. Thus, it provides second-level evidence for H.

Of course, in the overall assessment of H's probability one should also take into consideration the first-level evidence for it, i.e., H's probability conditional not just on ne, but on all relevant pieces of evidence e, e', e'', etc. However, as pointed out by Vickers, it is practically impossible to establish that a claim is future-proof just by assessing first-level evidence. In fact, especially because of the empirical underdetermination of theories, H's probability based only on first-level evidence cannot become sufficiently high. Instead, in ideal cases the NMA by itself may produce a reliable 95% consensus that a hypothesis describes a scientific fact.

There are problems, however, and not all cases are ideal. First, as explained above, in this way we cannot establish whether H was used essentially, hence whether it is completely true or not, and so whether it is actually future-proof. Alternatively, we might say that we can establish that it is future-proof *in the weaker sense* that *a part of it* is completely true, hence properly speaking future-proof. Second, the novel prediction ne may be not very risky, and the probability of H decreases as the logical probability of ne increases. Third, a NMA may be less than ideal also because it is unclear whether the prediction or predictions involved are actually novel, and even more because (as suggested by the literature on novel predictions), we should probably consider novelty as coming in degrees (Alai 2014b, p. 312).

Often, however, a hypothesis H is employed in more than one (more or less risky) novel predictions $ne_1, \ldots, ne_j$. For instance, Adams and Leverrier predicted both the position and the mass of Neptune and Mendeleev predicted various properties of different new elements. Of course, the conjunctive probability of many predictions gets smaller and smaller as the number j of predictions increases. Therefore, it can be very low, hence the probability that the hypothesis H from which those predictions were derived is true can be very high, even if each of those predictions is not *very* risky.

Instead, if it is unclear whether the predictions are actually novel, or if they are only partially novel, the probability of H will be proportionally lower.

Even this smaller probability, however, may raise the degree of confirmation conferred to H by first-level evidence.

Finally, the fact that H accounts for *old* evidence may also constitute additional second-level evidence for it. In fact, we may ask what the probability is of finding a hypothesis (whether true or false) which accommodates a known datum e. Apparently, the answer is that if the theoretician was minimally skillful, the probability was 1, for this was just a puzzle-solving exercise. But since there is an infinite number of false hypotheses for each true one, the probability that H is false is practically 1. However, things are seldom so simple. To begin with, H must usually account for *many* data $e_1, \ldots, e_m$, and the more they are, the harder the task becomes. Moreover, the data are practically never entailed *just* by H, but by H in conjunction with several collateral assumptions $A_1, \ldots, A_n$, which in turn must be derived from, or at least be consistent with, a number of independently accepted theories $T_1, \ldots, T_n$. So, one must also find the right $A_1, \ldots, A_n$, and H must not only entail $e_1, \ldots, e_m$, but be compatible with $T_1, \ldots, T_n$.

Therefore, finding a hypothesis entailing $e_1, \ldots, e_m$ (whether true or false) may become impossible for a minimally skilled theoretician, and very difficult even for a truly gifted one: the probability that H is false decreases with the number of accommodated data and of the collateral assumptions needed. If these numbers become very high, it is no longer plausible that H has been found by puzzle-solving skill alone, and another hypothesis becomes more plausible: that the theoretician was not just trying to accommodate $e_1, \ldots, e_m$, but, more importantly, looking for a true hypothesis using a reliable heuristic, so she actually found one.

This is why I once suggested that certain confirming instances which are apparently different from the confirmation provided by novel predictions are actually of a similar nature: for instance, the convergence of independent theories, the convergence of measurements by different experimental procedures based on independent theories, Keynes' distinction of confirming instances, and non-*ad hoc* explanations (Alai 2014b, § 4). In other words, this argument from the complexity of the theoretician's task can be turned into an argument from the improbability that just by chance independent theories converge in accounting for a large number of disparate data. Even this argument, however, may show at most that H is partly true, for even a partly false hypothesis may entail $e_1, \ldots, e_m$, and so be employed (inessentially) to account for them.

4 Conclusion

According to Vickers, a certain kind of second-level evidence (i.e., a 95% consensus) may show *that* certain claims are future-proof. Here I have suggested that certain different kinds of second-level evidence (provided by NMA-like considerations) may *justify* the achievement of such a consensus. That is, they can explain *why* and *how* it was reached for sound epistemological reasons rather than just for non-epistemic sociological drives, and hence why it can be a reliable indicator of future-proof claims. In other words, when at least 95% of specialists in a field take a claim as describing a scientific *fact*, they are probably right, at least in the *weak* sense that at least certain (possibly unidentified) parts of the claim are future-proof. A 95% consensus is justified especially when those claims yield various novel and risky predictions, but possibly also when they display great systematicity and unifying power (by accounting for a very large number of known data $e_1, \ldots, e_m$) and great plausibility and coherence with accepted theories and assumptions (by accounting for $e_1, \ldots, e_m$ in full coherence with assumptions $A_1, \ldots, A_n$ and theories $T_1, \ldots, T_n$).

Vickers grants that a claim fulfilling criterion (C) may be only *approximately* true, hence it may be future-proof *modulo-minor adjustments*. This qualification, however is a potential threat for his enterprise, for approximation is a vague concept, thus there is a continuum ranging from being simply true to being approximately true, half-way between true and false, more false than true, or, finally, completely false. So, if a 95% consensus can be achieved by claims that are only approximately true, why couldn't it be achieved by claims that approximate the truth less and less, or are even very distant from it? He suggests that in practice we can clearly distinguish when a claim is substantially true from when it is not. Perhaps we can, but only retrospectively: *ideally*, and with some approximation, we might suppose that 100% consensus shows that H is 100% true, 95% consensus shows that it is 95% true, and so on:

> The further a claim is from 100% true, the less likely it is that a truly solid consensus will be reachable. So, a claim only halfway true would probably never reach a 95% consensus. Other things being equal, there will be barriers to a halfway-true claim reaching 95% solid scientific consensuses that are not present for a claim that is true. E.g., it will likely be less thoroughly tested, and the scientific community will know that.[3]

However, things may be far from ideal, I am afraid there might be a very wide consensus on claims that are far from *completely* true (various examples are provided by the history of science). Again, therefore, it might be safer to assume simply that Vickers' criterion warrants that a claim is at least *partly* true, hence *weakly* future-proof.

Acknowledgment. I thank Peter Vickers for helpful comments on the first draft of this paper.

References

Alai, M. (2014a). "Defending Deployment Realism against Alleged Counterexamples", in G. Bonino, G. Jesson, J. Cumpa (eds.) *Defending Realism. Ontological and Epistemological Investigations*, De Gruyter, Boston, Berlin, Munich, pp. 265–290.

Alai, M. (2014b). "Novel Predictions and the No Miracle Argument", *Erkenntnis* 79, n. 2, pp. 297–326.

---
[3]Vickers in discussion.

Alai, M. (2014c). "Why Antirealists Can't Explain Success" in F. Bacchini, S. Caputo and M. Dell'Utri (eds.) *Metaphysics and Ontology Without Myths*, Cambridge Scholars Publishing, Newcastle upon Tyne pp. 48–66.

Alai, M. (2021). "The Historical Challenge to Realism and Essential Deployment", in *Contemporary Scientific Realism. The Challenge from the History of Science*, Lyons, T. and Vickers P. eds., Oxford University Press, New York 2021, pp. 183–215.

Alai, M. (2023). "Bayesian 'No Miracles Argument' and the Priors of Truth", preprint, Pittsburgh PhilSci Archive, https://philsci-archive.pitt.edu/23321/.

Cordero, A. (2001). "Realism and Underdetermination: Some Clues from the Practices-Up", *Philosophy of Science* LXVIII, pp. 301–312.

Cordero, A. (2017a). "Retention, truth-content and selective realism", in E. Agazzi (ed.) *Varieties of Scientific Realism*, Cham: Springer, pp. 245–256.

Cordero, A. (2017b) "Making Content Selective Realism the Best Realist Game in Town", *Quo Vadis Selective Scientific Realism?* Conference, Durham, UK, 5–7 August 2017.

Doppelt, G. (2007). "Reconstructing Scientific Realism to Rebut the Pessimistic Meta-induction". *Philosophy of Science*, 74: 96–118.

Doppelt, G. (2011). "From Standard Scientific Realism and Structural Realism to Best Current Theory Realism". *Journal for General Philosophy of Science*, 42: 295–316.

Kuhn, T. (1962). *The Structure of Scientific Revolutions*, The University of Chicago Press, Chicago, Ill., London.

Musgrave, A. (2006–7). "The 'Miracle Argument' for Scientific Realism". *The Rutherford Journal* 2. Article 020108.

Psillos, S. (1999). *Scientific Realism. How Science Tracks Truth*, Routledge, London.

Rescher, N. (1987). *Scientific Realism: A Critical Reappraisal*. Dordrecht: Reidel.

Stanford, P. K. (2006). *Exceeding Our Grasp: Science, History, and the Problem of Unconceived Alternatives*, Oxford: Oxford University Press.

Vickers, P. (2022). *Identifying Future-Proof Science*, Oxford University Press, Oxford.

White, R. (2003). "The epistemic advantage of prediction over accommodation". *Mind*, 112(448): 653–683.

Serendipity between psychology and logic of scientific discovery

Marco Buzzoni

Sezione di Filosofia e Science Umane, Università di Macerata, via Garibaldi 20, 62110 Macerata, Italy

Abstract. Serendipity is the phenomenon whereby a fortuitous and unexpected experience turns out to be an essential element leading to a scientific discovery or invention. The discussion of serendipity has led to the formulation of a "paradox of control": on the one hand, serendipitous discoveries are accidental and unpredictable, but on the other hand, they can be prepared and fostered. The paradox, already foreshadowed by Plato, brings to light the need to reconcile two essential elements of scientific discovery: unpredictability and genetic-methodological reconstructability. To resolve this paradox, it is appropriate to challenge both the acceptance of the Popperian (or neopositivist) distinction between psychology (or discovery) and logic (or justification) and its subsequent rejection within the epistemological tradition. This leads to a distinction between two senses—one reflexive, the other genetic-methodical—of the psychology/logic (or discovery/justification) dichotomy that resolves the paradox of serendipity. A critical analysis of Popper's considerations of accidental discoveries in science both clarifies more concretely the root of the paradox and to distinguish his eclectic solution from the one proposed here.

1 Introduction

Many philosophers of science have insisted on the complementarity, in science, of creative-subjective invention and methodological-objective justification. Karl Popper, for example, drew a well-known distinction between the psychology and the logic of knowledge, whose cooperation captures according to this author the very essence of scientific research. For him, science is characterized by two stages that, while in many logical senses opposite and chronologically distinct, are both necessary: the first characterized by an act of creative intuition, the second by the critical-methodical effort to check and falsify the products of that intuition.

A similar distinction is to be found in Henri Poincaré. He found in mathematics two entirely different kinds of minds: the "logicians" (*logiciens*) and the "intuitionalists" (*intuitifs*). The one group places logic in the foreground, leaving nothing to chance, the other group resorts first and foremost to intuition (Poincaré 1906, pp. 11–16, Engl. Transl. pp. 210–222). He, too, recognized the need for cooperation between these opposing attitudes of thought, emphasizing both the importance of preparation and accuracy in the formulation of a problem, in order to facilitate the next moment of creative invention, and the shortcomings of a purely logical-demonstrative procedure: "logic is not enough; [...] the science of demonstration is not all science and [...] intuition must retain its role as complement, I was about to say as counterpoise or as antidote of logic."[1]

Now, the problem underlying these and many other similar positions is that, on the one hand, creativity and method are two concepts that are both necessary for understanding scientific discovery, but, on the other hand, they, at least at first glance, seem mutually exclusive. It is precisely this complementarity and tension between the two concepts that has recently been taken up in the discussion around serendipity, i.e., the phenomenon in which a fortuitous and unexpected experience turns out to be an essential element leading to a scientific discovery. Within this discussion, in fact, a "paradox of control" has been formulated, according to which, on the one hand, serendipitous discoveries are accidental and unpredictable, but, on the other hand, they can be prepared and learned by an appropriate method. As we shall see, to resolve the tension between these concepts, it is necessary to rethink the relationship between creativity and method in scientific discovery or, more generally, the traditional distinction between psychology and logic, between the context of discovery and the context of justification. In this paper I shall try to resolve this paradox by showing that, by more adequately analyzing these pairs of concepts, it is possible to distinguish two senses—one reflexive (or transcendental), the other genetic-methodological—in which they can be understood. The two different points of view from which the concepts of creativity and method can be considered show that, far from being opposites or even antinomic, these concepts are complementary, such that each requires the other as its logical complement. For this purpose, however, creative invention and critical-methodical control should not be understood— as is the case of Popper (or of the logical empiricists or of Kuhn's endless cycle of normal and revolutionary phases)—as two separate components or phases of scientific research, which could exist and stand, as it were, separately side by side. Instead, the two concepts are never given separately from each other and can be distinguished only by counterfactual abstraction. Creative unpredictability and genetic-methodological controllability are two inseparable faces of the same concrete cognitive act. Creativity tends to resolve itself into the elaboration of particular scientific methods, which in turn redeem and transform the unpredictability (or "accidentality") of

---

[1] Poincaré 1906, p. 25; Engl. Translation, p. 35; on this point see also Poincaré 1908. Many later authors were inspired by Poincaré. G. Wallas, for example, was influenced by Poincaré in his proposal of the following four stages of creative thinking: "Preparation, Incubation, Illumination (and its accompaniments), and Verification" (cf. Wallas 1926, pp. 79–107). Poincaré's basic idea was also taken up by Campbell 1960, who interpreted it as favouring a blind-variation-and-selective-survival process for understanding all genuine increases in knowledge: cf. Campbell 1960, pp. 215–218 and 282–311.

the serendipitous event into a methodological path that is in principle reconstructible and intersubjectively controllable.

2   The concept of serendipity and the paradox of discovery

The English word "serendipity" was coined in 1754 by Horace Walpole on the basis of a fairy tale about "The Three Princes of Serendip" (the old name of the island of Ceylon), who, as he wrote in a letter to Horace Mann,

> were always making discoveries by accident or sagacity of things they were not in quest of: for instance, one of them discovered that a mule blind of the right eye had travelled the same road lately, because the grass was eaten only on the left side, where it was worse than on the right (Walpole 1754, pp. 407–408).

The most important milestone in the analysis of the concept is Merton and Barber's initially unpublished draft *The Travels and Adventures of Serendipity. A Study in Historical Semantics and the Sociology of Science*, dated 1958. Reworking and publication of this unpublished draft gave the final and decisive impulse to the fortune of the term in many fields of research, including that of the philosophy of science: first published in Italian translation in 2002, the work was followed two years later by the English edition in Merton and Barber 2004.[2]

Merton's two fundamental ingredients, unexpected and fortunate findings on the one hand and insight or wisdom on the other, return, with some variations, in almost all subsequent definitions. This applies not only to the definitions more often proposed within the epistemological debate (cf., e.g., Van Andel 1994, p. 643; Fine & Deegan 1996, pp. 434 and 445; McBirnie 2008, p. 604; Thagard 1998a and 1998b; Nickles 2009, pp. 179 ff.; Copeland 2019, p. 2386; Arfini, Bertolotti and Magnani 2020, p. 940 fn.), but also to those concerning more particular areas of research, such as information seeking, management, innovation, or recommender systems.[3] In fact, the number of works devoted to serendipity today seems to be increasingly concentrated in the latter areas (cf. Quy Khuc 2022), but in all of these

---

[2] Merton, however, had analyzed the concept of serendipity in works prior to the just mentioned draft. In his essay *Sociological Theory* (1945, p. 469n.), he already gave a concise definition of the phenomenon: "Fruitful empirical research not only tests theoretically derived hypotheses; it also originates new hypotheses. This might be termed the "serendipity component of research, i.e., the discovery, by chance or sagacity, of valid results which were not sought for" (Merton 1945, p. 469n.; the definition has been taken up both in Merton 1948, p. 506 and in Merton 1949, p. 98).

[3] For information seeking see e.g. Case 2007 (p. 337), Foster & Ellis 2014, and the important empirical study by Sun et al 2011. For management and innovation, see e.g. MacDonald 1998, De Rond 2005, Gherardi 2006, Fink et al 2017, and Busch 2022. For recommender systems, see e.g. Kotkov, Medlar and Glowacka 2023.

publications there is a clear need to first establish a general definition before embarking on specific investigations, which indirectly shows the need for a properly philosophical analysis of the concept, the sole object of this paper.

The main problem associated with the concept of serendipity is that the two main characteristics indicated by Merton are not easily reconciled and, indeed, that we are dealing with a paradoxical, if not oxymoronic or contradictory concept. The first to intuit this was Horace Mann, Walpole's correspondent. Mann not only directly relates the term "serendipity" to scientific research (which was not the case in the letter sent to him by Walpole), but also, indirectly, raises the problem we intend to discuss here, the tension between the accidentality and unpredictability of discovery on the one hand and the necessity of its intersubjective or methodological reconstructability on the other. Mann noted that the type of serendipity with which Walpole was endowed is very peculiar: not only does accidentality appear in it, but this accidentality is such that, once the "serendipitous" event is given, it universally leads to the discovery itself:

> I perfectly understand your 'serendipity'. It must have happened to everybody, that in searching for one thing, others of greater importance have occurred. How many useful discoveries, for example, has the search of the philosopher's stone produced, that the student was certainly not in quest of. Is not this 'serendipity'? But the *sortes Walpolianae* are still more useful, if you can find everything *a point nommé* whenever you dip for it. (Mann 1754, p. 415)

The problem raised by serendipity had emerged since the early days of philosophical thought. In 1994, Van Andel aptly called attention both to the paradoxicality of the concept of serendipity and to the relevance of well-known classical problems in the serendipity debate, choosing two significant exergues: a fragment of Heraclitus and Plato's eristic argument posed by Meno. The Heraclitus fragment is as follows:

> If you do not expect the unexpected, you will not find it; for it is hard to be searched out and difficult to compass.[4]

---

[4]Heraclitus, Fr. DK B 18, a fragment, however, which I quote from Marcovich's 1967 translation, p. 40. This translation seems preferable to me because it expresses the extreme difficulty, but not the impossibility, of finding the unexpected. Marcovich's translation, in fact, takes into account the fact that ἀνεξερεύνητον "mean[s] only 'hard to be searched out' and 'difficult to compass or discover', and not 'impossible to ...'." The Logos, the author points out, "is 'difficult to compass' either because it is hidden inside the things or because it is paradoxical" (Marcovich 1967, p. 40). Cf. also Kirk, Raven, and Schofield 1983, p. 193: "If one does not expect the unexpected one will not find it out, since it is not to be searched out, and difficult to compass." Cf. also Mason 2014, p. 68, fn. 18: "If one does not expect the unexpected one will not discover it, for it is not to be discovered and intractable".

Here the problem that is inherent to the concept of serendipity is formulated indirectly, with respect to the aim of discovering the ultimate essence of reality: does it make sense to expect the unexpected? According to Heraclitus, the answer is in the last analysis affirmative, though with an important *caveat*. However difficult to grasp, the *Logos*, which is the least familiar and least expected one can conceive, allows one to decipher, and thus expect, what is "indicated by signs", but not explicitly told, by the gods to men: the oracle in Delphi, in Heraclitus' own words, "neither speaks out nor conceals, but gives a sign." (Heraclitus Fr. DK B 93, transl. from Marcovich 1967, p. 51).

Later—as Van Andel 1994's second exergue rightly suggests—the problem will be re-proposed in the famous eristic argument formulated in Plato's Meno, according to which it is impossible for man to investigate both what s/he already knows and what s/he does not yet know:

> SOCRATES: I know what you want to say, Meno. [...] a man [...] cannot search for what he knows – since he knows it, there is no need to search – nor for what he does not know, for he does not know what to look for. (Meno 80 e; Engl. Transl. by G. M. A. Grube, in Plato 1997, p. 889)

If we leave aside the explicit use of the term "serendipity," it was in my opinion Thomas Nickles, in his numerous and always enlightening contributions concerning the concept of scientific discovery, who gave a striking formulation of the paradoxical character of serendipity. On the one hand, he wrote, "any method capable of generating interesting, new knowledge must incorporate an element of luck, chance, or contingency." (Nickles 2009, p. 179) But, on the other hand, "the idea that there could be a method of innovation based upon luck or chance or serendipity looks positively oxymoronic. Chance and luck are the very things that method traditionally is supposed to exclude." (Nickles 2009, p. 178)

As for finally the more recent discussion of this problem under the name of serendipity, Abigail McBirnie gave the most explicit formulation of a "paradox of control" inherent in the concept of "seeking serendipity": "[w]hile seeking serendipity seems improbable, paradoxically, some degree of control may be possible." (McBirnie 2008, p. 601) As the author explains, the paradox arises from the combination of, on the one hand, the "random, elusive and unpredictable nature" of serendipity, which seems to rule out any attempt to pursue it, and, on the other hand, method, "which suggests a purposive approach and a skill or ability that 'can be trained and encouraged'" (McBirnie 2008, p. 604).

How to solve the paradox of serendipity and, more generally, of discovery? The Platonic solution was only apparent or circular. The hypothesis according to which we bring back to memory something we have already known in a pre-birth life, when the mind's eye was not obscured by sensible appearances, merely shifts the problem from our embodied existence to that, even much less known, of a purely intellectual existence preceding our present, embodied one. How could we, in pre-birth life, have known new things (that is, "ideas")? What remains unexplained is precisely the possibility of discovery of new intelligible ideas of the *hyperouranios topos*.

But what about the solutions proposed in the serendipity debate? Sometimes they move in the right direction, but are not entirely satisfactory. Their most frequent flaw is that, instead of explaining at the root the coexistence of chance and method, they insist on the fact of this coexistence and make it plausible by resorting to concrete examples in which both elements are present. However, to simply insist that, despite the accidentality and unpredictability of serendipitous discovery, it is possible to foster unexpected discoveries is, in the final analysis, like refuting Zeno's arguments against the existence of movement by walking back and forth.[5]

Now, to outline how the seemingly opposite elements of serendipity can be conceived without falling into paradox, it is necessary to make a small detour, briefly addressing the problem that, in my view, lies at the heart of the paradox: the way of understanding the relationship (in Popper's lexicon) between the logic and the psychology of knowledge, or (in the lexicon of logical empiricists), the relationship between the context of discovery and the context of justification. For reasons of space, I will say only the minimum necessary to outline the solution of the serendipity paradox.

3 Two fundamental senses of the psychology/logic (and discovery/justification) distinction

The distinction between the psychology and the logic of knowledge is both one of the main pillars of Popper's philosophy of science and a point that, despite other differences, he shared essentially with the logical empiricist philosophy of science:

> I shall distinguish sharply between the process of conceiving a new idea, and the methods and results of examining it logically. [...]

---

[5]To this general claim (which applies above all to the essays oriented towards the search for concrete applications, as in the case of the literature focussing on information seeking or management) there are some notable, but partial exceptions, which would deserve a separate discussion, a task quite beyond the limits of this paper. See for example Nickles 2009, Catellin 2014, Arfini, Bertolotti, Magnani 2020; Glăveanu 2022, Copeland 2019, 2022, and 2023. These authors certainly move in the same direction of this paper. But there remains an important point of disagreement with them, which can be briefly summarized as follows: they do not draw a sufficiently neat distinction between the two fundamental senses in which, as we shall see, Popper's psychology/logic dichotomy (or the neopositivistic discovery/justification corresponding one) must be understood in order to resolve the serendipity paradox.

> [T]here is no such thing as a logical method of having new ideas, or a logical reconstruction of this process. [...] From a new idea, put up tentatively, and not yet justified in any way [...] conclusions are drawn by means of logical deduction. These conclusions are then compared with one another and with other relevant statements, so as to find what logical relations (such as equivalence, derivability, compatibility, or incompatibility) exist between them. (Popper 1935, pp. 4–6; quotations from the Engl. Transl., pp. 8–9)

In this way, Popper essentially took up the distinction discovery/justification that the logical empiricist philosophy of science had drawn (for historical details on this distinction, see Schickore and Steinle (eds) 2009, above all Part I and Part II, and Buzzoni 2015).

In general, the logical empiricists and Popper used the distinction to grant empirical science cognitive autonomy vis-à-vis the wider cultural and historical context. This was one of the reasons that the exponents of the relativistic philosophies of science of the 1960s (especially Kuhn and Feyerabend) and the advocates of the sociological turn (notably Bloor and Latour) from the 1980s onwards rejected the distinction in question. According to Kuhn and Feyerabend, for example, merely because they played an historical-causal role in the scientific process, *empirical-historical factors* such as scientists' prejudices and personal idiosyncrasies, aesthetic preferences, religious beliefs etc., are to be put on a par with more traditional *reasons* for maintaining or rejecting a theory, such as coherence, explanatory scope, unifying power, etc. (cf. Feyerabend 1970, § 14; Kuhn 1962, pp. 151–156; for an exponent of the sociological turn, see e.g. Bloor 1991, pp. 36–37).

In this way, however, the baby was thrown out with the bathwater. The baby was the minimal sense that I shall call here *reflexive-transcendental* (or simply *reflexive*) and in which reason is irreducible to empirical, particular causal factors, that is, as an expression of its claim to represent, in principle, things as they really are (no matter how far this can succeed). Although a countless number of physical, biological, psychological, sociological, and, generally, contingent or accidental factors influence and limit human reason, the irreducibility of this latter, at least in a sense, cannot be denied without denying all possibility of meaningful thinking or talking. Any claim to reduce reason to causal factors, necessarily presupposing its own truth, is irreducible to the causal factors to which, contradictorily, it grants a determining power over itself. In fact, to assert any empirical fact is to assert, implicitly, the distinction in principle between reason and facts, without which there would be neither one's own asserting nor one's own denying. At least in this sense the distinction between the contexts of justification and discovery is constitutive of reason and cannot be denied without contradiction, since it is affirmed by the very act of negating it.

So far I have defended the distinction in principle between psychology and logic of knowledge (or between context of justification and context of discovery) in the reflexive-transcendental sense, which expresses the irreducible autonomy of reason. However, we should distinguish at least another sense, which I shall call *genetic-methodological*, which is the opposite complementary of the reflexive-transcendental just seen, a sense in which this distinction must be entirely rejected.[6]

In fact, if the general claim of representing things as they are is not to remain devoid of any particular content and cognitive function, it must be realized by means of concrete methodological procedures which make it possible to reconstruct, to re-appropriate and to evaluate in the first person the reasons why a particular truth-claim should be accepted. In other words, the truth-claim of our discourses tends by its very nature – and more precisely as subordinate to the goal (in itself normative) of establishing itself as true – to translate (in principle without residue) into particular methods (or techniques).

Not only the logical empiricists, Popper and Lakatos, but also the exponents of the relativistic and sociological turn, failed to clearly identify this sense, in which a genetic-methodological (or genetic-historical) attitude is decisive for justification. To test the truth value of a statement, in principle we must always adopt this genetic and historical-reconstructive attitude and retrace the main methodological steps taken by those who first achieved a certain result through those steps. Pythagoras's Theorem can be used in a practical way without recalling the procedural steps of its demonstration. But if someone challenged its validity, we ought to test it by retracing in the first person the procedural steps that led to that theorem being asserted. By doing this, we *justify* a theory by historically reconstructing the context of its *discovery*. In this sense, context of discovery and context of justification are one and the same thing (for a more detailed justification of this thesis, see Buzzoni 1982 (ch. 3, § 1 and passim), 1986 (ch. 2 and *passim*), 2008 (ch. 1, §§ 4–7), and 2015).

4 Serendipity between psychology and logic of scientific discovery

The distinction between two senses – one reflexive, the other methodological (or genetic-methodological)—of the distinction between psychology and logic (or discovery and justification) of scientific knowledge allows to better understand and resolve the riddle of serendipity. As we have seen, on the

---

[6]Hoyningen-Huene 1987 carefully analyzes several senses of the discovery/justification distinction, but while these distinctions are certainly useful in particular contexts, none of them coincides with the one I have developed since Buzzoni 1982 and which is essential to defending the unity and distinction between creative invention and method in the sense of the central thesis of this paper.

one hand, if understood as an expression of the inescapable autonomy of the logical-discursive level of representation, the distinction between the logic and the psychology of knowledge must be maintained. The irreducibility of the rational value of our assertions about the world expresses the reflexive (or transcendental) dimension of discovery or invention, the ultimate source underlying all creativity, all emergence of what is new. This, on closer inspection, is also the point made by Popper against what he calls "historicism" by an argument that, though concerning the general growth of scientific knowledge, applies, *mutatis mutandis*, to the more limited growth of knowledge to be found in a single new discovery: "We cannot predict, by rational or scientific methods, the future growth of our scientific knowledge", since this would be tantamount to already knowing today what we will only know tomorrow (cf. Popper 1957, p. ix–x).

The argument holds that one cannot suppress the character of unpredictability (and/or chance, in a sense still to be clarified) that accompanies not only discoveries usually considered paradigmatic of serendipity, but any discovery as such. Every new idea has two sides, distinct but inextricably connected: on the one hand, as an expression of our rationality in its reflexive-transcendental sense examined above, it is, in a purely formal sense, an absolute beginning that cannot be reduced to actual causal factors. However, each new idea, while not predictable, can be satisfactorily explained ex post, after it has materialised in particular contents, drawn from experience or the world of culture.

In other words, even if from the transcendental point of view rationality is in principle absolute and free from conditioning, it is nevertheless, and indeed precisely because of this, entirely conditioned on the side of content. As we have seen, the inescapable sense in which human reason claims its autonomy is specifically realized in the process of knowledge through a particular set of methods, that is, etymologically, of retraceable "ways" or "paths," without the indication of which the fundamental scientific value of intersubjective controllability would be lost. This is the case because, however formally absolute, scientific creativity is in principle nevertheless subordinated to the personal commitment of the scientist to witness how things are in themselves, seeking to bracket any subjective biases or idiosyncrasies towards the object. Now, concretely, this commitment is realized precisely in putting in place a series of methods or procedures that, in principle, must be traceable and reconstructible by anyone in the first person and must lead to the ascertainment of actual courses of events independent of our subjective will.

This stems from the complementarity between the reflexive level of creativity and the genetic-methodological level of the principled reconstructability of any scientific discovery. Finding a question that the event we have stumbled upon provides an answer to, is not something that could have been predicted (at least in its determinacy) before it actually occurred—which is why we consider it a "creative" performance of the human mind—but, in light of the preceding considerations, it must be to some extent reconstructible (and thus to some extent comprehensible and predictable) after the discovery has occurred, and precisely on the basis of the methodical steps that, starting from the initial accidental event, led—and can in principle lead again any agent endowed with mind and body—to the discovery itself. Serendipity, considered in this light, is a particular example of a general phenomenon, which consists in the possibility of inserting *any* actual event already happened into a rational-explanatory discourse, i.e., one endowed with intelligible and intersubjectively testable meaning. What was previously accidental and fortuitous for us disappears as such and becomes a step in the genetic reconstruction of the demonstrative-experimental procedure that led to the discovery.

To better explain this last point, the account of thought experiments I have developed elsewhere proves to be an important aid. According to my account, one of the most general conditions of the possibility of formulating thought experiments lies in the typical capacity of human reason to transform any data or empirical circumstance into something that is hypothetically counterfactual, and only insofar as it can be thought or imagined as such, it can be inserted into the meaningful whole of our discourses, and more generally, into the meaningful whole of human culture. The ability to give new meaning to facts already known from experience, placing them in a new counterfactual context, is ultimately the same capacity that underlies our ability to experiment in thought.

This is true of the simplest perceptions. I am only able to perceive the red of a rose because I can hypothetically and counterfactually assume the possibility that it is of any other colour, and then reject this possibility on the basis of my empirical perceptions. Even a declarative sentence like "the sun is shining" has meaning only against the background of the possibility that the sun might not be shining. This sentence expresses an empirical observation that is the answer to a cognitive question concerning a hypothesis about the state of the sun; without this hypothesis, which usually remains in the background and is not explicitly addressed, the observation that the sun is shining would have no definite meaning. But this is also true in general. To be able to formulate thought experiments is the condition of possibility to conceive of, and then execute, real world experiments (for more details on this point, see Buzzoni 2008, pp. 115–116).

In the capacity to imagine things as something different from what they actually are lies the first condition of the unpredictability of human discoveries: we cannot place an a priori limit on finding new and different ways of looking at reality. In its properly transcendental sense, the distinction between the rational context of justification and the historical context of discovery is not only irreducible (in the sense that the rational value of our assertions is irreducible to any set of historical factors or circumstances), but also allows us to grasp the transcendental value of discovery or invention in itself, the true nature of the creativity that underlies every emergence of what is new. The first condition for an accidental event to be included in the conceptual path of a discovery is that it is assumed to be a purely hypothetical or counterfactual event. Without this, both simple observations and real experiments would be, strictly speaking, unintelligible. Tackling the problem under this perspective, the mind's ability to imagine counterfactual scenarios, the ability to see things differently from how they actually are, makes any experience, including "accidental" ones, a plausible answer to hypothetical questions that we are able to formulate. From this point of view, events that are accidental (and as such not only unplanned, but also independent of us) become parts of a thought experiment, which may lead to a scientific discovery. Empirical discoveries always move from contingent conditions and end with the formulation of some new question to which those conditions, now transformed into parts of the counterfactual scenario of a thought experiment, can be regarded not as the first, but the last elements of a chain of events that provides an intersubjectively reproducible and therefore testable answer.

This is more than the usual claim of the unpredictability and freedom of scientific research, so far as we are in a position to avoid the risk of assuming only one of the two fundamental aspects of serendipity (the reflexive-transcendental one), and neglecting the other (the genetic-methodological). The mentioned risk is avoided from the outset by the complementarity – the key concept in our explanation of serendipity – of the reflexive-transcendental level of creativity and the genetic-methodological level of the intersubjective testability in principle of every concept. As already seen, if the reflexive-transcendental claim to represent, in principle, things as they really are is not to remain devoid of any particular content and cognitive function, it must be realized by means of genetic-methodological procedures, which make it possible to genetically reconstruct, to re-appropriate and to evaluate in the first person the reasons why a particular truth-claim should be accepted.

Some examples can illustrate what we have been saying. Consider first a simple example taken from everyday life. I notice a stone in my path. Is it simply an obstacle, because it might be something I might stumble upon? Or does its shape suggest to me (perhaps because of some affordances in Gibson's sense) that I can use it as a scraper to sharpen other tools? In one sense it is certainly true that the answer will certainly depend, in its specificity, on my prior "background knowledge" (Popper), acquired habits, etc. But, if one does not neglect the properly reflexive-transcendental side of the problem, this will always also depend on the human capacity to interpret what one sees (a stone in my path) as a plausible answer to a hypothetical question that has arisen in the course of our interaction with the world around us (for example: "How could I make sharper arrowheads?").

Thus, finding a question to which the event we stumbled upon provides an answer is not something that could have been foreseen at a time before it happened, both because it concerns a real event independent of human will, and because our explanation was not univocally predetermined a priori. Nevertheless it must now be reproducible (and therefore comprehensible and predictable) after the discovery has taken place, and precisely on the basis of the methodological steps that, starting from the initial accidental event, have led – and can in principle lead – to the discovery itself.

Now, in the same way, we must treat the cases most clearly related to serendipity. Note first that the same facts that for other people were purely accidental—and therefore inexplicable—for the three Principles of Serendipity were the logical and at the same time practically reconstructible consequences of their reasoning. But let us look at historically real examples. Take for instance Fleming's discovery of penicillin, one of the most cited and investigated examples of serendipity. As well known, Fleming observed that in a culture plate, accidentally contaminated by a mould, the bacterial growth of Staphylococcus colonies was inhibited. Many accidental factors favoured the discovery, which – as has been noted – were due to several concomitant circumstances, some of which had an exceptionally low probability of occurrence even when taken in isolation: the poor tidiness in the laboratory, very particular bacteria that had colonised the Petri dish, the weather conditions, and many others as well (cf. Waller 2002, pp. 251–255). However, if we reconstructed in detail what happened to Fleming from the first fortuitous co-occurrences to his discovery, we would find exactly what he found. Even the initial accidental events are no longer pure coincidences, but the initial moments of a mental and at the same time practical-experimental pathway that we can still retrace now. As we can see, the principle of genetic-methodological reconstructability is also respected in this kind of discovery, which is paradigmatic of serendipity.

The same applies to the classic Newtonian apple, which became an example of the (hypothetical) law of universal gravitation. The first fall was accidental (both in the sense of being a real event independent of human will and in the sense of not being foreseeable on the basis of the knowledge of the time), but the place it later occupies, both in Newton's first explanation and in the explanation we can give today, is well determined. Or take Semmelweis's discovery of the cause of puerperal ("childbed") fever: nursing mothers who were in the ward and therefore not accessible to doctors who had previously handled corpses did not fall ill by pure chance. Since the aetiology of puerperal fever had not yet been discovered, there was no reason why some women fell ill and others did not. But this case, when translated into genetic-methodological rationality, became in principle a technically reproducible effect, generating the rule not to visit women who had given birth without first washing hands thoroughly. Also the initial, accidental and contingent moments of discovery can always, in principle, be reconstructed rationally.

In all these cases, from the new point of view, an event formerly serendipitous (and inexplicable except by means of the mind's freedom to construct possible counterfactual courses of events), is now no longer so, unless we move our minds into the past, before it took place, before the mind constructed a counterfactual course from which it is now possible to represent in the mind a series of events ending with the very fact from which the mind itself had started in its explanatory effort.

In the next section, we shall briefly examine the way in which Karl R. Popper, without using the term serendipity, addressed the role of chance in scientific discovery. As I shall try to show, it is precisely Popper's inability to grasp the sense in which it is necessary to distinguish, alongside the transcendental-reflexive sense of human reason, a genetic-methodological sense, which prevented him from going beyond an eclectic position, not essentially better than the psychological considerations of Pasteur.

5 Popper and accidental discoveries in science

Like other authors, Popper insists that, in spite of its unexpected and accidental character, every discovery presupposes a mind prepared to exploit this chance. As is well known, according to Popper, no empirical finding can count as a discovery if it does not acquire its meaning from the point of view of a theoretical expectation that we seek to refute (and which is part of a background knowledge, without which research could not advance one step). From this point of view, and bearing in mind Popper's equation between the information content of a theory and its improbability, it is no surprise that Popper provided *ante litteram* a relatively simple answer to the problem of the serendipitous character of scientific discoveries.

According to the fundamental methodological rule of Popper's falsificationism, we ought to test our most cherished theoretical hypotheses in order to falsify them, and we can only accept them as corroborated if this corroboration somehow surprises us and makes us see that, even if we thought our hypothesis was false, against all probability it has withstood our best checks. According to Popper, in fact, the empirical content of a scientific theory is the greater the more improbable the theory is: the more a scientific hypothesis says about the real world, the greater its empirical content, the greater the number of its potential falsifiers, i.e., the imaginable circumstances in which it could be falsified. Or, correlatively, the more a scientific hypothesis says about the real world, the less likely it is to be corroborated. Here, the (partially) accidental nature of corroboration is reconciled, at least at first sight, with the need to have both a hypothesis that makes sense for future observations and all the experimental preparation necessary to put the theory to the test:

> Lavoisier's experiments were carefully thought out; but even most so-called 'chance-discoveries' are fundamentally of the same logical structure. For these so-called 'chance-discoveries' are as a rule refutations of theories which were consciously or unconsciously held: they are made when some of our expectations (based upon these theories) are unexpectedly disappointed. Thus the catalytic property of mercury was discovered when it was accidentally found that in its presence a chemical reaction had been speeded up which had not been expected to be influenced by mercury. But neither Örsted's [sic!] nor Röntgen's nor Becquerel's nor Fleming's discoveries were really accidental, even though they had accidental components: every one of these men was searching for an effect of the kind he found. We can even say that some discoveries, such as Columbus' discovery of America, corroborate one theory (of the spherical earth) while refuting at the same time another (the theory of the size of the earth, and with it, of the nearest way to India); and that they were chance-discoveries to the extent to which they contradicted all expectations, and were not consciously undertaken as tests of those theories which they refuted. (Popper 1969, pp. 220–221).

Popper's answer is ultimately a version of Pasteur's oft-cited comment on Hans Christian Ørsted's discovery of the "electric telegraph": "in the fields of observation chance favours only prepared minds (*esprits préparés*)" (Pasteur 1854, p. 131). The accidentality of discovery is greatly attenuated by the scientist's degree of "preparation" or already acquired knowledge, which also attenuates the concomitant phenomenon of surprise, whereby, as Aristotle already noted, what surprises the ignorant does not surprise the knowledgeable (cf. *Met.* 983a 12–20).

The scientist's previous background knowledge, in fact, makes it possible to give theoretical meaning to observations that would otherwise remain devoid of any cognitive significance, even if, Popper adds, it must be admitted that neither Ørsted's nor Röntgen's nor Becquerel's nor Fleming's discoveries "were really accidental, *even though they had accidental components*".

Now, on closer inspection, Popper continually confuses the two senses in which it is possible to understand both the ('psychological') unpredictability and the ('logical') reconstructability of scientific discoveries.

On the one hand, he confuses two meanings of unpredictability: one is that which concerns the accidental event which, in hindsight, favours discovery, but which is in itself a real event independent of our will; the other is that of discovery as a mental representation that makes a real event at first sight completely isolated fit into the web of causal relations already conceptualised. Now, while the latter sense is transcendental and cannot be predicted (as Popper rightly argued, we cannot have knowledge today of what we will know/discover tomorrow), the former is accidental and unpredictable only because of its independence of our subjective will. Strictly speaking, in fact, what is accidental in the sense of being unpredictable for us (in the light of our current knowledge) is, at least from a heuristic perspective, representable as potentially part of a determinate causal chain, which in fact, in the light of the discovery made, will be methodologically perfectly reconstructible.

In other words, any event that occurs without it being possible to indicate a reason why it occurs or does not occur is by definition accidental (like the number on the wheel of fortune or the number that comes up after rolling a dice), but as we discover other circumstances or conditions under which it occurred, it becomes more and more probable and our ignorance diminishes: it approaches asymptotically, often without ever being able to reach it, the limit value of a deterministic event. Lightning, which for primitives was so indeterminate as to be attributed to the capricious will of God, is today not only (largely, not entirely) predictable but, under certain experimental conditions, even reproducible in the laboratory.

In this case, however, the most important significance of accidentality is that it presupposes the existence of an external reality that is independent of our subjective will as far as its content is concerned. In this sense, accidentality is by no means paradoxical, but, on the contrary, is a condition of possibility of science as an enquiry concerning an independent reality. Nevertheless, this accidentality must be 'redeemed' in a different sense: the accidental event must be turned into an initial situation from which follow a series of steps leading, in an intersubjectively testable manner, to its explanation, i.e., to a discovery.

In addition to confusing two distinct concepts of the accidentality of scientific discovery, Popper also confuses two distinct concepts of the hypothetical moment. Popper (like Pasteur, Leibniz, and many other recent authors) is certainly right to argue that the first condition for an accidental event to be included in the conceptual path of a discovery is that it be thought of in the light of some particular hypothesis. But this very capacity to formulate particular hypotheses depends on the more radical capacity in general of the mind on which we insisted above. Unlike in Popper, this radically hypothetical character of science cannot be confused with the particular hypotheses that are formulated from time to time to understand our experience. Instead, the individual concrete hypotheses developed for this purpose are the way in which the reflexive or transcendental side of serendipity is translated into concrete genetic-methodological steps. They only embody the methodical aspect of discovery, without which its reproducibility, its intersubjective controllability, in short its scientificity, would be lost.

In addition to particular hypotheses, one must distinguish the ability as such to counterfactually assume a hypothetical horizon that defines the space of meaning as such. The simplest observation of what reality is like presupposes that it can be hypothetically different. As already mentioned, I can perceive the red of the rose I am looking at only because I can hypothetically assume the possibility that it might have a different colour (and I can then reject this possibility on the basis of my empirical perceptions). This possibility, as a general possibility or capacity to formulate hypotheses (and not as a specific hypothesis), ultimately coincides with the mind's capacity to give cognitive meaning to the perceptual-real datum.

Popper's solution ultimately remains eclectic because of the confusion between the transcendental-reflexive meaning and the genetic-methodological meaning of the fundamental concepts he uses. More precisely, it is a double confusion because it concerns both the moment of accidentality as it relates to the moment of the "psychology of knowledge" and the moment of methodological reconstructability as it relates to the moment of the 'logic of knowledge'. Because of this double confusion, for example, he treats accidentality as an element, an ingredient or, as he literally puts it, a "component" of scientific discovery. He says that neither Ørsted's nor Röntgen's nor Becquerel's nor Fleming's discoveries "were really accidental, even though they had accidental components". Now, how is it possible not to be accidental and yet have accidental components? In fact, as we know, these discoveries were accidental in the sole sense of being unpredictable, but they were not accidental at all insofar as a prepared mind was able to recognize an event which, however anticipated in the mind, was itself wholly determined in the causal chains that existed independently of the subjective will of the scientist.

But the main difficulty Popper runs into, due to the above-mentioned untraced distinctions, is even more serious and repeats in essence the untenability of his distinction between psychology and research logic. According to Popper, the "accidental component" sometimes comes to the aid of other components, that is, the expectations generated by previously given hypotheses (the preparation Pasteur spoke of), but how this can happen again and again in the course of scientific research remains ultimately something quite inexplicable. It is of course true that serendipitous discoveries necessarily presuppose hypothetical or theoretical antecedent assumptions that prepare or open the minds of researchers, but one does not really dispel the paradox unless one provides a properly philosophical explanation of this happy cooperation and correspondence between hypothetical anticipation and empirical discovery. How is it possible for the researcher to have formed in his or her mind precisely those theoretical assumptions or expectations that will later find a correspondence—accidental and therefore improbable by its very nature—in reality? Until this (or other similar and interrelated) questions are satisfactorily answered, it will not be possible to avoid a certain eclecticism, which rightly requires the combination of chance and method, but is unable to clarify the conditions of its possibility.

It is therefore by no means a coincidence, but the consequence of a fundamental distinction that has been overlooked, that for Popper serendipity can be no more than the simple eclectic sum, combination or juxtaposition of "planned insight coupled with unplanned events", to use Fine & Deegan's general definition of serendipity (cf. 1996, p. 445). In reality, as we have tried to clarify, every discovery originates from accidental (because not understood and therefore a fortiori unplanned) events that only thanks to the fundamental capacity of the human mind to assume anything real as hypothetical, become in the last a part of a "planned insight", i.e., a particular hypothesis that can represent a law-like concatenation of real events independent of our subjective will.

6 On the distinction between serendipitous and non-serendipitous discoveries

A consequence of what we have said is that, strictly speaking, every discovery is characterized by a certain degree of serendipity. From this point of view, the difference between discoveries that are serendipitous and those that are not can only be one of degree. This is in no way to deny that the historian or sociologist or psychologist may deem it appropriate to establish a certain boundary between cases of discovery in which the characteristics of serendipity are particularly marked. But however much one wants to insist on this difference in degree, one should not turn it into a qualitative difference.

This was clear to Merton. He rightly tried to better circumscribe the concept of serendipitous discovery, writing that what he called the "serendipity pattern" refers to "the fairly common experience of observing an unanticipated, anomalous and strategic datum which becomes the occasion for developing a new theory or for extending an existing theory." (Merton 1949, p. 98; but see also the more detailed interpretation of these terms on pp. 98–99.)

Some have sought also qualitative criteria. One of the most interesting attempts that seems to me to move at least to some extent in this direction is that of Arfini et al. 2020. According to the authors of this essay, serendipity phenomena occur not when a discovery was "wildly" unexpected, but when it was "reasonably" unexpected. For example, the invention of the post-it note, which is clearly a kind of serendipitous discovery, took place because

> the chemists and engineers involved were glue-experts, and were able to recognize it. It made sense to them, it did fit the knowledge they had and the projections about their ignorance, so they were able to understand it. Had they stumbled upon something radically different, such as something with no gluing power but an amazing strawberry smell, they would have probably shrugged and thrown the batch away. It would have been something so wildly unexpected that it would have been uselessly bewildering. (Arfini et al 2020, p. 943)

The question now arises whether the distinction between a "wildly" unexpected discovery and a "reasonably" one is to be understood as a distinction of degree or of principle. In the former case we would be in the presence of an interesting characteristic, which in many contexts can be usefully added to those already specified by Merton.

If, on the other hand, the distinction were understood as a kind distinction, the criterion on the basis of which it is drawn would have to be rejected. In fact, because of the context in which the distinction in question is placed and on the basis of the example that illustrates it, it seems to be proposed as a principled distinction between serendipitous discoveries and discoveries without scientific relevance. Now, it is true that, in the sense we have called reflexive (or transcendental), the distinction marked by the terms "reasonably" and "wildly" would be qualitative, but as such it distinguishes true scientific discoveries from *entirely* accidental events or *entirely* irrelevant hypotheses. From this point of view, strictly speaking, there can be no accidental discoveries without scientific significance, but only mere events to which, at a certain stage of scientific research or human culture, we have not yet been able to attribute any cognitive significance.

It does not help much to resort to the distinction, proposed by Hendricks and Faye 1999, of two different types of abduction: "paradigmatic" and "trans-paradigmatic". Although useful for the specific purposes of particular historical-empirical research, even this distinction remains, strictly speaking, only a difference of degree and not, as the Arfini et al 2020 seem inclined to assume, a difference of principle. According to these authors, "paradigmatic" abduction is connected to discoveries that play the role of "game-changers" in scientific progress and therefore are genuine cases of serendipity (cf. Arfini et al. 2020, p. 946).

It is clear that the plausibility of the distinction between "paradigmatic" and "transparadigmatic" abduction is closely related to Kuhn's distinction between normal and revolutionary science. But while it is true that the distinction between evolutionary and revolutionary stages of scientific change may play an important role in the history of science (perhaps justified by the particular purposes it serves from time to time), it cannot be interpreted as a qualitative difference, especially in the case we are discussing: neither from extraordinary science nor from 'normal' science can the note of creativity, novelty, unpredictability be completely expunged.

If the distinction between serendipitous and non-serendipitous discoveries is to remain a distinction of degree, in fact characterizing only the most serendipitous discoveries from those that are less so, it is good to reiterate one last time that the principled distinction that should not be overlooked is that between what is discovery for us and what is merely an actual event, since this distinction expresses the irreducible autonomy of human reason. From this point of view, however, all discoveries are serendipitous, both because they are all the result of human reason and because they are all marked by some degree of accidentality. For this reason, even the solution of a "puzzle" in Kuhn's sense is not guaranteed. It too may be unexpected at some point (for example, if so many good researchers in the field in the past have failed to find a solution) and requires a certain amount of creativity.[7]

7 Conclusion

In the context of the discussion on serendipity, the "paradox of control" emerged: on the one hand, serendipitous discoveries are accidental and unpredictable, but, on the other hand, they can be prepared, fostered and learned. The paradox, already anticipated by Heraclitus and Plato as a paradox related to obtaining new knowledge, brings to light the need to reconcile two essential elements of scientific discovery: unpredictability and genetic-methodological reconstructability. To resolve this paradox, I questioned both Popper's acceptance of the distinction between the psychology and the logic of knowledge and his later rejection of it as part of the relativistic-sociological turn.

This led to a distinction between two senses—one reflexive, the other genetic-methodical—of the psychology/logic (or discovery/justification) dichotomy. This distinction makes it possible to resolve the paradoxicality of serendipity (and scientific discovery in general) by clarifying in what precise sense both lucky initial chance and unpredictable human discovery are reconcilable with the principle of intersubjective testability of all scientific knowledge. A critical analysis of Popper's considerations of accidental discoveries in science both clarified the solution proposed here of the paradox of control and helped to capture the eclectic nature of Popper's position.

Acknowledgements. Different versions of this paper were presented at the workshop "Imagination, Thought Experiment and Embodiment in Scientific

---

[7]I developed this objection in Buzzoni 1986, pp. 32–51. Toulmin (1972, p. 106–107) was among the first to stress the need to understand the distinction between normal and revolutionary science as one of degree.

Knowledge" (Macerata, November 2022), and at the international congress "Model-Based Reasoning in Science and Technology. Inferences & Models in Science, Logic, Language, and Technology" (Rome, June 2023). I am grateful for several helpful criticisms and suggestions provided by the members of the audience of these conferences, especially Mike Stuart, Hans Jörg Rheinberger, Selene Arfini, Aliseda Atocha, Gerhard Heinzmann, Lorenzo Magnani. The final version of this work has also profited from a research stay at the Max-Planck-Institut (Berlin, April 15 to May 30, 2023) sponsored by the Alexander von Humboldt Foundation, to which I am sincerely grateful.

References

Arfini S., Bertolotti T., Magnani L. 2020. The Antinomies of Serendipity How to Cognitively Frame Serendipity for Scientific Discoveries. Topoi 39, 939–948.

Bloor D. 1991. Knowledge and social imagery. 2nd ed, Routledge and Kegan Paul, London

Busch C. 2022. Towards a Theory of Serendipity: A Systematic Review and Conceptualization. Journal of Management Studies, 1–50 (doi:10.1111/joms.12890).

Buzzoni M. 1982. Conoscenza e realtà in K. R. Popper. Angeli, Milan.

Buzzoni M. 1986. Semantica, ontologia ed ermeneutica della conoscenza scientifica. Saggio su T.S. Kuhn. Angeli, Milan.

Buzzoni M. 2008. Thought Experiment in the Natural Sciences. Würzburg, Königshausen+Neumann.

Buzzoni M. 2015 The Practice Turn in Philosophy of Science: The Discovery/Justification Distinction, and the Social Dimension of Scientific Objectivity, pp. 81–111. In E. Agazzi and G. Heinzmann (eds.), The practical Turn in Philosophy of Science, Angeli, Milano.

Campbell D.T. 1960. Blind Variation and Selective Survival as a General Strategy in Knowledge Processes. In: M. C. Yovits and S. Cameron (eds.), Self-organizing Systems, 205–231. Pergamon Press, New York.

Case D. 2007. Looking for Information, 2nd ed., Academic Press, London.

Catellin S. 2014. Sérendipité. Du conte au concept. Seuil, Paris

Copeland S. 2019. On serendipity in science: Discovery at the intersection of chance and wisdom. Synthese, 196(6), 2385–2406.

Copeland S 2022. Metis and the Art of Serendipity. In Ross W. and Copeland S. (eds) 2022. The Art of Serendipity, pp. 41–73. Macmillan, Palgrave.

Copeland S. 2023. Serendipity and the History of the Philosophy of Science. In: S. Copeland, W Ross & M. Sandet (eds), Serendipity Science, pp. 101–123. Cham.Springer.

De Rond M. 2014. The structure of serendipity. Culture and Organization, 20(5), 342–358.

Feyerabend P. K. 1970. Against method: Outline of an Anarchistic Theory of Knowledge. In: Radner M, Winokur S (eds), Analyses of Theories and Methods of Physics and Psychology, Minnesota Studies in the Philosophy of Science, IV. University of Minnesota Press, Minneapolis, pp. 17–130.

Fine G. A., and Deegan J.G. 1996. Three principles of Serendip: Insight, chance, and discovery in qualitative research. International Journal of Qualitative Studies in Education, 9(4), pp. 434–447.

Fink, T. M. A., Reeves, M., Palma, R., & Farr, R. S. (2017). Serendipity and strategy in rapid innovation. Nature communications, 8(1), 1–9.

Foster, A. E., & Ellis, D. 2014. Serendipity and its study. Journal of Documentation, 70(6), 1015–1038 (doi: 10.1108/JD-03-2014-0053).

Gherardi S. 2006. Organizational Knowledge: The Texture of Workplace Learning, Blackwell Publishing, Oxford.

Glăveanu Vlad P. 2022. What's 'Inside' the Prepared Mind? Not Things, but Relations. In Ross W. and Copeland S. (eds) 2022. The Art of Serendipity, pp. 23–39. Macmillan, Palgrave.

Hendricks FV, Faye J 1999. Abducting explanation. In: Magnani L., Nersessian N. J., Thagard P. (eds) Model-based reasoning in scientific discovery. Kluwer Academic/Plenum Publishers, New York, pp. 271–294.

Hoyningen-Huene P. 1987. Context of discovery and context of justification. Studies in History and Philosophy of Science Part A, 18(4), pp. 501–515.

Kirk G. S., Raven J. E., Schofield M. 1983. The Presocratic Philosophers. A Critical History with a Selection of Texts, 2nd ed, Cambridge University Press, Cambridge/New York/New Rochelle

Kotkov D., Medlar A., and Glowacka D. 2023. Rethinking Serendipity in Recommender Systems. In ACM SIGIR Conference on Human Information Interaction and Retrieval (CHEER '23), March 19–23,

2023, Austin, TX USA. ACM, New York, NY, USA, 5 pages (doi: 10.1145/3576840.3578310).

Kuhn T.S 1962. The Structure of Scientific Revolutions, University of Chicago Press, Chicago (quotations are from the second edition, 1970)

MacDonald S 1998. Information for Innovation, Oxford University Press, Oxford.

Mann H. 1754. Letter from Mann, Friday 8 March 1754. Quotations are from W. S. Lewis, W. H. Smith and G. L. Lam (eds.), Horace Walpole's Correspondence with Sir Horace Mann, Vol. 4, pp. 414–416. Yale University Press, London 1960.

Marcovich, M. 1967. Heraclitus. The Greek text with a short commentary. Los Andes University Press, Merida.

Mason, Andrew J. 2014. Heraclitus' Usage of ὅστις in Fragments DK B 5 and B 27. Phronimon 15(2), pp. 55–68.

McBirnie A. 2008. Seeking serendipity: the paradox of control. Aslib Proceedings: New Information Perspectives, 60(6), pp. 600–618.

Merton, R. K. 1945. Sociological Theory. American Journal of Sociology, 50, pp. 462–473.

Merton, R. K. 1948. The Bearing of Empirical Research upon the Development of Social Theory. American Sociological Review, 13, No. 5, pp. 505–515.

Merton, R. K. 1949. Social Theory and Social Structure. Toward the Codification of Theory and Research. The Free Press of Glencoe, Illinois.

Merton, R. K. and Barber, E. 2004. The Travels and Adventures of Serendipity. Princeton University Press, Princeton.

Nickles T. 2009. The Strange Story of Scientific Method. In: J. Meheus and T. Nickles (eds.), Models of Discovery and Creativity, pp. 167–207. Springer, Dordrecht Heidelberg London New York.

Pasteur L. 1854. Discours à l'occasion de l'installation solennelle de la Faculté des lettres de Douai et de la Faculté des sciences de Lille. Impr. A. d'Aubers, Douai. Repr. and quoted from: Oeuvres complètes réunies par Pasteur Vallery-Radot. Tome VII: Mélanges scientifiques et littéraires, pp. 129–132. Masson et Cie Éditeurs, Paris. 1939.

Plato 1997. Complete Works, Edited, with Introduction and Notes, by John M. Cooper, Hackett Publishing Company, Indianapolis/Cambridge.

Poincaré H. 1906. La valeur de la science, Flammarion, Paris. Engl. Transl. by G. B. Halsted, The Foundations of Science, Science Press, New York. 1913, pp. 201–358.

Poincaré H. 1908. L'invention mathématique, Bulletin de L'institut Général Psychologique, 1908, Vol. 8, 175–187. Engl. Transl. by G. B. Halsted, Mathematical Creation. The Monist, 1910, Vol. 20 (3), pp. 321–335.

Popper K. R. 1935. Logik der Forschung, Springer, Wien. Engl. Transl. From The Logic of Scientific Discovery. Routledge, London, 2002.

Popper K. R. 1957. The Poverty of Historicism. The Beacon Press, London.

Popper K. R. 1969. Conjectures & Refutations, Routledge & Kegan Paul, London, 3rd revised edition.

Quy Khuc 2022. How do we perceive serendipity? In: QH Vuong (ed), A New Theory of Serendipity: Nature, Emergence and Mechanism, pp. 13–40. De Gruyter, Berlin.

Schickore J. and Steinle F., eds. (2006) Revisiting discovery and justification: historical and philosophical perspectives on the context distinction. Springer, Dordrecht

Sun, X., Sharples, S., & Makri, S. (2011). A user-centred mobile diary study approach to understanding serendipity in information research. Information Research, 16(3).

Thagard P. 1998a. Ulcers and bacteria I: discovery and acceptance. Stud Hist Philos Sci C 29(1):107–136.

Thagard P. 1998b. Ulcers and bacteria II: Instruments, experiments, and social interactions. Studies in History and Philosophy of Science Part C: Studies in History and Philosophy of Biological and Biomedical Sciences, 29(2), 317–342.

Toulmin S. 1972. Human Understanding. The Collective Use and Evolution of Concepts, vol. I, Princeton University Press, Princeton (N.J.).

Van Andel P. 1994. Anatomy of the unsought finding. Serendipity: Origin, history, domains, traditions, appearances, patterns and programmability. Br J Philos Sci 45(2): 631–648.

Wallas G. 1926. The Art of Thought. Cape, London.

Waller J. 2002. Fabulous Science: Fact and Fiction in the History of Scientific Discovery. Oxford University Press, Oxford.

Walpole H. 1754. Letter to Sir Horace Mann, Monday 28 January 1754. Quotations are from W. S. Lewis, W. H. Smith and G. L. Lam (eds.), Horace Walpole's Correspondence with Sir Horace Mann, Vol. 4, Yale University Press, London, 1960, pp. 407–411.

Realism, scientific creativity, and theory justification

Alberto Cordero

CUNY Graduate Center, 365 Fifth Avenue, New York NY 10016, United States of America & Philosophy Department, Queens College, City University of New York, 65-30 Kissena Blvd, Flushing NY 11367, United States of America

> **Abstract.** Scientific realists generally interpret novel empirical success and scientific fecundity as indicators that at least some of the assumed theoretical content is true. However, an influential anti-realist argument, revived by Kyle Stanford (2015, 2019, 2021), challenges this 'conservative' expectation. This presentation discusses the argument and concludes that its premises do not apply to methodologically reflexive versions of selective scientific realism.

1 Scientific realism

I will focus on realist positions emphasizing the epistemic value of novel empirical success and scientific fecundity (disclosure of new phenomena or previously unnoticed relationships between already known phenomena). That stress has a broad following in science. As Ernan McMullin (1984) noted at the start of the contemporary debate, "The near-invincible belief of scientists is that we come to discover more and more of the entities of which the world is composed through the constructs around which scientific theory is built." To McMullin's generation of realists, the expected benefit of the approach was to explain theoretical success in ways that reveal general indicators we can use to select parts of empirically successful theories that offer persistent epistemic achievements in science. Thinkers as varied as Putnam, McMullin, and Maddy, among others, linked the truth of empirical theories to empirical success. With clarifications and modifications, this expectation remains firm in contemporary projects. Realists argue that when theories show strong empirical success, it is reasonable to attribute the success to a systematic relationship or connection between the theory's representation of how things are in a certain part of the world and that part of the world. We thus have the following realist thesis:

> **Thesis R°:** A hypothesis's empirical success and fertility indicate that at least some of the theoretical ideas it assumes are true.

R° rests on daily experience and the history of theories methodologically focused on novel prediction. In many branches of science, successions of theories commonly exhibit retentions (effective if not exact) of theoretical parts, many of which "persist" robustly. This phenomenon is particularly apparent in the modern natural sciences. One sterling example is the retention of classical mechanical equations found as special cases in ordinary quantum mechanics, an outcome strengthened in recent decades by realist efforts to track classical ontology to the quantum mechanical evolution of the quantum state under particular regimes of scale and energy. (See, e.g., Wallace 2012, and section 5 below).

Why do elucidations of the effective working-level emergence of earlier ontologies matter to the noted thinkers? Realism postulates a link between empirical success and truth content, one allegedly strong enough for successful theories to give us more than mathematical constructs. Hence, there is an expectation that successful theories, even when destined to be superseded, yield correct representations of at least part of the unobservable world to which they refer—ones that somehow survive theory change. Importantly, in scientific practice, theoretical retention is not just a philosophical idea but a practical reality in the natural sciences. For instance, despite the stark ontological differences between General Relativity and Newtonian theory, they share working-level (functional) descriptions of specific physical regimes. This agreement gives us some understanding of why Newton's gravitational theory remains successful under specific energy and scale conditions, demonstrating the continuity and evolution of scientific theories. To realists, the details of the intertheoretical relationships between successor and predecessor descriptions *explain* why the latter work so well under specifiable regimes. From their perspective, it is in this "working/effective" sense that radical theoretical change is compatible with the truth of selected parts of the earlier theory.

In the opposite interpretive camp, non-realist thinkers emphasize the existence of a multitude of historical episodes of radical revolution at the level of theoretical foundations. In response, realists attentive to the last half-century of philosophical analyses of the history of science recognize the need to compromise. One concession they make is the expectation of new drastic discontinuities in scientific ideas to come. Successful empirical theories are usually wrong about some aspects of their intended domains. As whole constructs, empirical theories are probably false. However, reformed realists stress that a false theory may contain truthful parts, pointing out that, despite the recurrence of radical conceptual change, there are substantial continuities between the dominant theories in the classical period and the contemporary ones. This emphasis on conservation of content is the hallmark of "Selective Realism," a family of projects variously developed in the last decades of the previous century, especially by John Worrall (1989), Philip Kitcher (1993), and Stathis Psillos (1999), and subsequently furthered by, e.g., Mario Ala (2021), Alberto Cordero (2017), Matthias Egg (2016, 2017), and Peter Vickers (2019), among many others. The new selective projects moderate the traditional realist expectations but maintain the idea that diachronic inter-theoretical relations will continue to support truth attributions for selected parts, especially in the case of theories rich in corroborated novel predictions.

The presumption of theoretical content conservation has critics who think selectivists often oversimplify existing historical counterexamples. Here are two influential objections: (1) the proposed criteria for selecting theoretical parts are seriously defective, and (2) the expectation of retention of theoretical parts tends to weaken the imagination and bias the planning of future research. Objection (1) has a point. Many of the selection criteria proposed have allowed for unfortunate choices. For example, Saatsi and Vickers (2011) point to seriously incorrect theoretical components that, they claim, have played a crucial role in generating scientific successes, for instance, the luminiferous aether and Kirchhoff's diffraction theory. As noted, ongoing responses to these warnings include working-level (functional) selective approaches developed in recent years. The resulting projects seem promising, but the debate remains in full swing.

With this background in mind, let us move to my main topic, not objection (1) but an antirealist argument within (2) that allegedly devalues theoretical retention epistemologically and methodologically. It's important to note that retention is a feature central to working-level realism and other reformed projects of selective realism.

2 A tempting argument

The argument I wish to discuss builds on a complaint revived in recent years by Kyle Stanford in "hard" (2015) and "softer" (2019, 2021) versions that seek to exhibit the scientifically impoverishing character of realist commitment. The argument has two central premises.

> P1: Realist commitment to successful theories encourages skepticism towards proposals incompatible with the commitments adopted.
>
> P2: In contrast, not being committed to theoretical content makes scientists systematically more open to radical novelty and correspondingly more creative—with more modest convictions than those of "committed" scientists, but also better justified.

The conclusion is that realist commitment limits scientific imagination and creativity counterproductively. Stanford believes this shows how realism and antirealism differ regarding how we should plan scientific investigations. In a similar vein, Brement (2007) argues that, at least concerning successful theories, realists tend not to see the need for what funding agencies like the NSF call "transformative science" (e.g., the discovery of metallic glasses) and "revolutionary disciplines" (e.g., plate tectonics), or the need to create entirely new fields or disturb established theories. According to Stanford

(2015), even the most tolerant realists tend to react suspiciously to research projects that contradict the scientific theories' elements, aspects, or features to which they are committed. Realists act like this—he affirms—because they believe they have an epistemic basis to favor research that preserves "well-established" content at the expense of revisionist theories.

For non-realists, the practical impact of the retention thesis differs significantly from that for realists. Stanford stresses that realists have reasons that non-realists lack for disfavoring proposals that violate existing theoretical orthodoxy. As a result, realists tend to be more satisfied than non-realists with evaluation committees that reject theoretical proposals that contradict current theories. He suggests that non-realists are more open to theories that challenge their own without being willing to accept any theory. For instance, constructive empiricists limit their belief to the empirical consequences of the best scientific theories, casting doubt on proposals that contradict the best-established observable consequences of received theories. They are, however, open to promoting proposals that challenge the parts least accessible to observation in current theories (e.g., about the nature of dark matter).

3 Backing up the argument

Stanford's thesis is initially plausible. Realist commitment has encouraged disregard for the evidence and derision of alternative approaches in the past, while a lack of theoretical preference has benefited the exploration of novel theoretical frameworks. These observations fit with numerous scientific episodes, highlighting the potential relevance of the thesis.

(a) Many illustrative episodes are from when the sciences operated under supposed "undeniable truths." For instance, the traditional conception of uniform circular motion in astronomy as the natural motion of heavenly bodies; the doctrine of natural places in pre-modern physics and biology (including ideas of rigid natural hierarchies in human groups, e.g., men and women); and the Cartesian conception of ontological dependence in wave physics, among many other ideas. We now appreciate that many of them hindered the scientific imagination, a fact that should intrigue and pique the curiosity of scientific realists. Analogously, in pre-Darwinian biology, there is the approach of natural theology, according to which complex systems in nature show the existence of intelligent design in the world, a view compellingly articulated by the Reverend William Paley (1746–1805)[1]. According to Paley,

---

[1]As Paley put it: *"Suppose I [found] a watch upon the ground, and it should be inquired how the watch happened to be in that place [...] When we come to inspect the watch, we perceive [...] that its several parts are framed and put together for a purpose, e.g. that they are so formed and adjusted as to produce motion, and that motion so regulated as to point out the hour of the day; that, if the different parts had been differently shaped from what they are, of a different size from what they are, or placed after any other manner, or in any other order, than that in which they are placed, either no motion at all would*

the human eye provided an incontestable example of nature's purpose and design toward perfection. It proved, he thought, the existence of a Designer. In principle, the eye could have developed cumulatively at random, as David Hume had already admitted in his *Dialogues concerning natural religion*. But such gradual aggregation required the availability of an indefinitely long time, against all imaginable expectations then. Until the mid-nineteenth century, intelligent design seemed the only conceivable explanation.

Paley's work is an exemplar of realist natural philosophy. It discouraged exploring anti-teleological ideas in biology until, in the second half of the 19th century, discoveries about the character and scope of spontaneous change shadowed some of the dearest intellectual intuitions that had sustained biology for millennia. In keeping with Stanford's thesis, the ensuing revisions were primarily the work of empiricist thinkers—some moderate, like Darwin, and others radical, like Mach. However, a significant shift was brewing in the empiricist camp. And with it, a new era of open-mindedness was dawning in science, a change that would revolutionize our understanding of the world. Einstein's Special Theory of Relativity is a testament to this new mindset. Historical studies suggest that for Einstein and other scientists at the turn of the century, the winning philosophy was neither "anti-realism" nor realism but an explicit fallibilist new scientific realism, a trait reflected in the subsequent epistemologies, most influentially Karl Popper's (see, e.g., Howard 1993).

(b) Open-mindedness was not universally practiced, however (it isn't now). Blocking theories contrary to orthodoxy did not end with the devaluation of a priori intuitions at the beginning of the 20th century. An instance in point is the conservative blockade practiced against geological mobilism during the central part of the last century. Mobilists were reacting to *fixism*, a long-entrenched conception according to which the continental crust and ocean basins are stable (fixed). Mobilism claimed that continents undergo large-scale lateral movements, drifting through the seafloor and forming a more significant landmass. While some physical indications supported mobilism, the geological establishment rejected continental drift. Objectors argued that there was no proper evidence for continental movement, no feasible mechanism for it, and no predictable patterns to the proposed movements. They branded the theory as "immature" (Giller et al., 2004; Doppelt, 2007). Opposition to mobilism remained strong until the 1970s. System-

---

have been carried on in the machine [...] There must have existed, at some time, and at some place or other, an artificer or artificers, who formed [the watch] for the purpose which we find it actually to answer; who comprehended its construction, and designed its use. (...) Every indication of contrivance, every manifestation of design, which existed in the watch, exists in the works of nature; with the difference, on the side of nature, of being greater or more, and that in a degree which exceeds all computation." [*Natural Theology* (1802)] atic discrimination against mobilist proposals, it seems, fueled intransigent adherence to fixist positions (Gradowski 2022).

Nevertheless, as Gradowski points out, there was qualitative evidence for mobilism at the time. It included

1. the geographical complementary fit of the continents, recognized since the 16th century,

2. cross-continental fossils of the same extinct land species,

3. continuities and geographical correspondences in geomorphological and stratigraphic data,

4. Paleomagnetic data in which sets of nearby magnetic rocks recorded vastly different locations of the magnetic poles upon their cooling, and discrepancies between continental and seafloor radiometric data indicated that the seafloor was relatively young.

However, the case against mobilism had merit. The evidence mobilists used was open to multiple interpretations, allowing for various consistent views. Additionally, fixist theorists argued that mobilist theory lacked coherence, adding another layer of complexity to the debate. On the other hand, fixist invoked an array of ad hoc land bridges connecting the two continents to account for the fossil evidence that suggested the same species had lived on the now vastly separated coasts of eastern South America and western Africa—as many bridges as necessary to save the appearances (Bryson 2004, Chapter 12). So, fixist explanations were not better.

For present purposes, the case is one of many examples illustrating the dangers of realist overconfidence in mainstream scientific research (see, e.g., Gradowski 2024). However, we must note a relevant difference regarding the suggested danger over the last century. The realist stance in science has developed projects of greater sophistication, and institutional science has gained methodological refinement. While individual scientists still sometimes take unwarranted stances, there is a noticeable shift in scientific communities generally favor more reflective stances (a significant difference from earlier times).

With the above background in mind, let us now discuss the suggestion that fallibilist positions of selective realism tend to hinder rather than help scientific originality and creativity compared to instrumentalist or non-realist positions. I will deny that Stanford's argument applies to the more reflexive versions of contemporary realism.

4 The premises

I start with P2, the noted argument's second premise, according to which the ability to articulate radically novel theories benefits from not having theoretical commitments:

P2: not being committed to theoretical content makes scientists systematically more open to radical novelty, and correspondingly more creative—with more modest convictions than those of "committed" scientists, but also better justified.

An old objection to P2 stresses the intellectual stagnation encouraged by non-realist and instrumentalist positions in diverse areas. Critics of radical empiricism have repeatedly made this complaint over the past century. Popper lamented that instrumentalist representations omit "the universe of realities behind the various apparencies" (1962: 8–40). According to W.B. Bonnor (1958), for radical empiricists, prediction is the full extent of a theory's importance, belittling the fact that many theories have revolutionized our perspective on space, time, matter, and life. More categorically, Nicholas Rescher (1987) says, "In foresaking realism, we would lose any prospect of developing a naturalistic account of why the phenomena are as they are. And this is too great a price to pay. A weighty argument against skeptical instrumentalism is that it immediately blocks any prospect of explaining why the phenomena are as they are—an explanation that must, in the nature of things, itself proceed in ultimately non-phenomenal terms" (1987, Chapter Four). These critics complain that antirealist interpretations of science impoverish the theoretical quest. In their view, realist ontological narratives fertilize theories and scientific imagination.

A key question is: Do the more reflective projects of realism tend to impoverish the imagination, leading to scientific stagnation? Realists answer in the negative, pointing to episodes like Einstein's research on Brownian motion, a phenomenon that was explained by the kinetic theory of matter, leading to the argument for the existence of atoms and molecules, the development of the geology of plate tectonics, and numerous fruitful corroborations of Darwinian stories, among myriads of similarly guided achievements. Realists further note the absence of compelling evidence for the alleged systematic fostering of creativity and discoverability by anti-realist stances, as claimed. In particular, directing science toward empirical adequacy at the expense of ontological realism has been tried many times. Still, it has not consistently led to more creative insights or better-justified theoretical narratives. By contrast, realists stress, from Galileo to Einstein and then to the present, significant advances in theoretical physics have benefited from incorporating thought experiments that fly above the observable world. These experiments are not blind guesses but apply theoretical narratives to a hypothetical situation and explore the possible world in which said situation is real, deducing consequences from the proposed scenario. The guesswork involved is often remarkable for the ability to produce ideas (creativity) and imagination (ability to transform ideas into reality) they exhibit.

Let us turn now to premise Q1. Compared to non-realist positions, does realism discourage self-criticism, imagination, creativity, or the justification of theoretical descriptions? If so, how? What adverse effects follow from a realistic stance on a theory or selected parts? On the face of it, contemporary realist projects promote opening the mind to new possibilities. However, more than this preliminary observation is needed to suggest that Stanford et al.'s anti-conservative position lacks scientific evidence. As we have seen, many scientific advances have been prevented, delayed, or derailed by assumptions of achieved knowledge. Realists cannot combat Stanford's thesis simply by declaring it intrinsically counter-scientific.

Here is a more promising starting point. To vindicate the interest in theoretical content retention shared by realists, we can begin by noting how science has changed epistemologically since the end of the 19th century and how the changes impact the realist project. Contemporary disciplines generally embody fallibilism and avoid closing the mind to previously unexpected possibilities. They discourage the epistemic overconfidence displayed in previous centuries. Several relevant developments are in view. There is more philosophical awareness at ground-level science than before. And, in the philosophy of science, realist positions have gained considerable sophistication. A brief detour on these developments is in order.

5 Some relevant features of scientific theorizing today

(1) While individual scientists still display obstinate conservatism sometimes, attitudes at the communal level have grown fairer. Epistemological and methodological awareness have improved, prompted by knowledge gained over the last century. Scientific communities are now more aware of the epistemological limitations of their work, and philosophers are more appreciative of the scientific background to their ideas. Current realist positions generally incorporate fallibilism, naturalism, and selectivism.

(2) In numerous episodes of theorizing, realist commitment ostensively leads to feats of creativity and improved justification of the theories involved. One representative example is the rise of ontic theories in quantum mechanics, particularly the main proposals associated with David Bohm, Hugh Everett, and objective quantum state collapse theories. Initiated in the 1950s to seek alternatives to the anti-realist interpretations promoted by the then dominant "Copenhagen Interpretation," at least three proposals have developed considerably since the 1980s. All of them are realist projects that take the quantum state as a physical state (Brown 2019) and the vindication of classical physics in specific scale and energy regimes. Despite

their conservative nature, I argue that the mentioned theories can be termed "progressive" due to the originality and the fruitful openness of scientific imagination and creativity that they exhibit, inspiring further exploration and advancement in the field of quantum mechanics.

5.1 The Copenhagen interpretation

The "Copenhagen Interpretation" (CI) was a family of theories united by a core of radical ideas that functioned as the official guide to quantum physics until the 1970s. It postulated drastic limits to the intelligibility sought by physics. According to the most radical empiricist versions, (a) the physical world possesses only those properties that direct experience reveals, and (b) accepting a theory means only believing in what the theory says about observable things and events in the world and not in any hypothetical reality that may or may not lie beneath appearances.

On the positive side, CI demonstrated remarkable scientific fruitfulness in many physical applications, from quantum mechanics to quantum field theory, significantly advancing our understanding of the physical world. On the downside, however, CI's limits on intelligibility seemed arbitrary. A very dark aspect was the ontological status accorded to measurement processes. Instead of explaining what happens when physical systems enter into measurement situations, CI declared it "analyzable," giving it only "black box" representations through a quantum algorithm that glossed over the processes' detailed physical description. In the 1930s, Einstein and numerous physicists and thinkers declared this restriction gratuitous. It was not acceptable, they argued, to have anything like it in something presented as the most basic physical explanation of material systems. These critics saw a commitment to obscurantism in CI. In 1935, their intellectual discontent gained detailed expression in an argument formulated by Einstein, Podolsky, and Rosen (EPR argument), a significant milestone in the history of scientific critique. In now historical discussions with Niels Bohr (the patriarch of CI), Einstein and several physicists offered realist arguments to refute CI through thought experiments like the one presented in the EPR argument. However, their efforts were not convincing enough, and the controversy became "metaphysical," remaining in that state for decades.

5.2 Three ontic theories

Insurrection against CI revived in the 1950s, led by the development of intellectually more ambitious theories such as 1952's David Bohm's Mechanics and 1957's Hugh Everett's Many Worlds Theory. In the following decade, 1964's Bell's Theorem hinted at ways of empirically deciding whether nature fully follows the classical principles of determinacy, separability, and locality. Soon, experiments based on generalizations of Bell's theorem began to tilt the epistemological balance toward quantum mechanics against both classical metaphysics and the radical empiricist strictures of CI. Maintaining the three principles mentioned seemed impossible—at least one had to be set aside. The revival of interest in the foundations of quantum mechanics, particularly ontology, was encouraged from various directions, notably experimental results on quantum interference and diffraction, arguments from partial absorption experiments (e.g., in single neutron interferometry), fruitful explanations of the stability of ordinary matter, and more (Harvey Brown 2019). Crucially, in these efforts, the winning ontology is not classical physics. The quantum state seemed fundamentally incompatible with classical expectations in all the theories mentioned, presenting a significant challenge and complexity that realists needed to address.

One point of interest here is that ad hoc assumptions, lack of clarity, and conceptual incoherence hopelessly marred all the initial versions of the ontic theories. Nevertheless, critical revisions led to significant improvements in the respective projects. Since the 1990s, three direct descendants of the approaches have dominated the realistic rebellion: Bohmian mechanics, the Many Decohering Worlds Quantum Mechanics, and spontaneous collapse theories—for example, those developed by Giancarlo Ghirardi and his collaborators in the 1980s (see, e.g., Cordero 2011 and 2019). In the revised theories, ad hoc assumptions give way to theoretical derivations from arguably reasonable models of initial conditions (e.g., Valentini 1991). The leading proposals naturally recover the descriptions of classical mechanics in particular regimes in the quantum domain. Recent versions of the many worlds approach or "multiverse" significantly improve probabilistic discourse (David Wallace 2012). In the case of spontaneous collapse theories, the tension between stochastic change of quantum state and relativistic physics is reduced (e.g., Philip Pearle 2000). These achievements of imagination and internal coherence, which had seemed impossible a decade earlier, are truly inspiring. Recall, for example, Hilary Putnam's principled Rejection of the Many Worlds Approach in the 2000s because he could see no way for it to yield meaningful probabilities (Meir Hemmo and Itamar Pitowsky, 2007).

The ontological proposals mentioned are complex and describe different physical worlds, each making divergent predictions. This divergence enables us, in principle, to choose between them in the laboratory. Unfortunately, the disagreements occur in areas that are (and may long remain) empirically inaccessible, adding another layer of complexity to our understanding. The ongoing debate about the ontological proposals is engaging, as it prompts us to ask: is any of the proposals more convincing than the others? None wins in predictive power—all are 'effectively' equivalent. The proposed theories differ, however, concerning other virtues, mainly simplicity, epistemic modesty, range of application, fertility, and explanatory power, keeping the debate alive and engaging. These differences translate into divergent selections of the "best option" (Cordero 2001, Callender 2020). However, while comparing ontological proposals is a fascinating issue, space limitations force me to stick to our central theme here—the profound impact of realist projects on the opening of the human imagination.

The development of the three ontic theories has expanded the scientific imagination beyond what was thought possible, particularly in the field of quantum science. Current interpretations of Everett's project show how to think of identity, individuality, and separability within the multiverse of the quantum world. Quantum state collapse theories, in turn, suggest ways to reconcile, at working (functional) levels, descriptions from general relativity and quantum mechanical "counterparts" invoking chance and discontinuous transitions. These reformed proposals have significantly improved the justification of the approaches, primarily through the effective compatibilization of descriptions provided by disciplines that had seemed impossible to integrate at any level, like classical and quantum mechanics. In this way, realist projects have helped break down barriers that held back imagination and creativity, inspiring new ways of thinking. Analogous developments are apparent in many other scientific areas, notably in fundamental physics, chemistry, biology, and psychology. All the noted improvements overshadow Stanford's premises against content retention. The final section elaborates on this idea.

6 The scientific internalization of realism

I have suggested that Stanford's premises against ontological engagement underestimate the creativity of realist projects like the ontic theories highlighted in the previous section. A second complaint concerns the inapplicability of the premises to more sophisticated versions of contemporary scientific realism. The latter has significantly transformed since the 1960s when naive ambition guided the prevailing realist projects. As Robert Klee (1999 313–4) recalls, a widespread belief at the time was that "our mature scientific theories, the ones we use to ground our scientific projects and experiments, are mostly correct" and "the errors they contain are minor errors of detail." Today, virtually no informed realist is so bold. While there are still instances of individual scientists embracing hard-nosed realist views, the community has shifted towards more moderate views tied to stringent conditions on evidence. This shift in community views is a significant development in the field, impacting the perspectives of philosophers of realist persuasion and the direction of scientific naturalization projects initiated in the 1980s by Dudley Shapere, Ernan McMullin, Ronald Giere, and Kitcher and in recent decades by a host of selective realists.

I use the term "naturalization" methodologically, focusing on Shapere's (1984) view that it is science itself that, in its fallible ways, identifies the relevant factors for discussing the ends, scope, and limits of knowledge. In this epistemological option, the philosophical analyses and conclusions spring from reasons internal to scientific activity. They do so in the form of specific considerations (as opposed to global or metaphysical ones) that are scientifically successful and free of reasonable doubt (i.e., well-founded). All conclusions are open to revision in light of new reasons and discoveries—there is no room for absolute trust. This version of realism confines epistemic commitment to just those parts of theories tentatively deemed well-founded by extant public standards. The credible parts are those specifically invoked to articulate predictions that prove correct, not the whole theory. From this perspective, the realist significance of corroborated predictions of theory parts is underscored by the systematic and varied predictive success that grounds the realist claim here: the theory parts invoked in the derivation of initially improbable corroborated predictions have non-trivial truth content. The resulting realist stances, all broadly empiricist, are fallibilist and reject *ideological* conservatism (of the sort that discriminated against mobilist theories in the 1950s). Criteria of coherence and novel empirical support are crucial in strongly constraining the acceptance of a theoretical idea.

Bringing these considerations home, a key point against Stanford et al.'s anti-conservative argument is the significant role of fallibilism in preventing conservative excesses in the empirical sciences over the last century. In the more alert projects of naturalized realism, the awareness of fallibility, fortified by the criterion that theories without strong novel empirical backing have no place within the realist stance, acts as a safeguard against conservative excesses. I have suggested the case of ontic quantum mechanical theories as exemplars of projects attentive to the need to remain open-minded about the state of knowledge. The case suggests how, in reflective contemporary disciplines, realist commitment can (and often does) promote scientific imagination and creativity, thereby enhancing the quality of scientific discourse, while limiting the scope of its claims and improving the justification of realist proposals.

A final observation here is that making ontological commitments may or may not systematically foster scientific creativity or the justification of ontological commitments. I have challenged only the allegation of systematic connections suggested by Stanford et al. Scientific creativity and justification navigate a sea of ever-changing contingencies. My point is that adopting ontological commitments does not lead to a systematic impoverishment of imagination or the search for justification. The opposite outcome, where adopting ontological commitments leads to an enrichment of imagination and the search for justification, seems more frequent in many disciplines.

References

Alai, Mario (2021): "The Historical Challenge to Realism and Essential Deployment In: Contemporary Scientific Realism." In: *Contemporary Realism*. Edited by: Timothy D. Lyons and Peter Vickers, Oxford University Press. Oxford University Press 2021: 184–215. DOI: 10.1093/oso/9780190946814.003.0009.

Bement, A. L. Jr. (2007) Important notice 130: Transformative research, National Science Foundation, Office of the Director.

Bonnor, W. B. (1958): Instrumentalism and Relativity." *The British Journal for the Philosophy of Science* (Vol. 8): 291–294.

Brown, Harvey R. (2019): "The Reality of the Wavefunction: Old Arguments and New". In Cordero A. (Editor), *Philosophers Look at Quantum Mechanics*. Cham: Springer (Synthese Library): 63–86.

Callender, Craig. (2020): Can we quarantine the quantum blight? In S. French and J. Saatsi (Eds.), *Scientific Realism and the Quantum*. Oxford University Press: 57–77.

Cordero, Alberto (2017): "Retention, Truth-Content and Selective Realism. In Evandro Agazzi (ed.), *Scientific Realism: Objectivity and Truth in Science*. Cham: Springer Nature: 245–256.

Doppelt, G. (2007): "Reconstructing Scientific Realism to Rebut the Pessimistic Meta-induction". *Philosophy of Science* (74):96–118 (2007).

Giller, P. S., Myers, A. A., and Riddle, B. R. (2004): "Earth history, variance, and dispersal." In M. V. Lomolino, D. F. Sax, & J. H. Brown (Eds.), *Foundations of biogeography: Classic papers with commentaries* (pp. 267–276). Chicago: The University of Chicago Press.

Egg, Matthias (2017): The Physical Salience of Non-Fundamental Local Beables." *Studies in History and Philosophy of Modern Physics* 57 (2017): 104–110.

Egg, Matthias (2021): "Quantum Ontology without Speculation", *European Journal for Philosophy of Science* 11 (2021): 32.

Giere, Ronald N. (1984): "Justifying Scientific Theories", in *Understanding Scientific Reasoning*, 2nd edition. Rinehart & Winston, Fort Worth: 96–110.

Gradowski, Laura (2022): *Facing the Fringe*. Graduate Center, CUNY: Ph.D. Dissertation.

Gradowski, Laura (2024): "From fringe to mainstream: The Garcia effect." *Studies in History and Philosophy of Science* (103): 114–122.

Hemmo, Meir; Pitowsky, Itamar (June 2007). "Quantum probability and many worlds." *Studies in History and Philosophy of Science Part B* (38): 333-–350.

Howard, D. (1993). Was Einstein Really a Realist? *Perspectives on Science* (1): 204–251

Kitcher, Philip (1993): *The Advancement of Science*. Oxford: Oxford University Press.

Klee, Robert (1999): "Realism and Antirealism." In R. Klee (Editor), *Scientific Inquiry: Readings in the Philosophy of Science*. New York: Oxford University Press: 313–16.

McMullin, Ernan (1984): "A Case for Scientific Realism; McMullin" In J. Leplin (ed.), *Scientific Realism*. University of California (1984): 8–40.

Pearl, Philip (2000): "Wavefunction Collapse and Conservation Laws". *Foundations of Physics* (30): 1145–1160.

Popper, Karl (1962): *Conjectures and Refutations*. New York: Harper Torchbooks.

Psillos, Stathis (1999): *Scientific Realism*. London: Routledge.

Rescher, Nicholas (1987): *Scientific Realism*, Chapter Four ("Against Instrumentalism: Realism and the Task of Science." D. Reidel Publishing Company. Dordrecht, Holland: 10–25.

Saatsi, Juha and Peter Vickers (2011): "Miraculous Success? Inconsistency and Untruth in Kirchhoff's Diffraction Theory." *British Journal for the Philosophy of Science* (62): 29–46.

Shapere, D. (1984): *Reason and the Search for Knowledge*. Dordrecht: Reidel.

Stanford, Kyle (2015): "Catastrophism, Uniformitarianism, and a Realism Debate That Makes a Difference." *Philosophy of Science* 82: 867–878. 21.

Stanford, Kyle (2019): "Unconceived Alternatives and Conservatism in Science: The Impact of Professionalization, Peer-Review, and Big Science." *Synthese* (196): 3915–3932. DOI: 10.1007/s11229-015-0856-4.

Stanford, Kyle (2021): "Realism, Instrumentalism, Particularism: A Middle Path Forward in the Scientific Realism Debate", in Tim Lyons and Peter Vickers, eds., *Contemporary Scientific Realism: The Challenge from the History of Science*, Ch. 10, Oxford University Press.

Valentini, A. (1991:, "Signal Locality, Uncertainty, and the Subquantum H-Theorem", Part I in *Physics Letters* A 156: 5–11; Part II in *Physics Letters* A 158: 1–8.

Vickers, Peter (2019): "Towards a Realistic Success-to-Truth Inference for Scientific Realism. *Synthese* (196): 571–585.

Wallace, David (2012): *The Emergent Multiverse: Quantum Theory according to the Everett Interpretation*. Oxford: Oxford University Press.

Worrall, J. (1989): "Structural Realism: The Best of Both Worlds?" *Dialectica* (43): 99–124.

Plausible hypothesis constructed by abduction: some examples in sciences

Jean-Pierre Desclés

Sens, Texte, Informatique, Histoire (STIH), Université de Paris Sorbonne, 1 rue Victor Cousin, 75005 Paris, France

Abstract. The notion of abduction (with the meaning given by C. S. Peirce) is essential for the formation of new knowledge. However, it has not received enough attention from the philosophers of science. The abductive process runs in different domains of science. In astronomy, the discovery of movements of planets around the sun has been imagined by an abductive process, against Tycho Brahe's system. According to Georges Polya, abduction is also very important in mathematics, it is a creative in this field of sciences. In linguistics, it is by abductive inferences that Champollion discovered and understood the system of hieroglyphs of Egyptian old documents and that Ferdinand de Saussure discovered some phonemes of an (non observed and hypothetical) Indo-European Language.

The notion of abduction (or retroduction) introduced by C. S. Peirce (*Collected Papers*)[1], also defended by George Polya[2] under the name of "heuristic syllogism", is essential for the formation of new knowledge; however, it has not received enough attention from the philosophy of sciences; often misunderstood and misinterpreted, abduction has not acquired an adequate place in the study of the creativity in scientific activities. I have already underlined its importance in other papers[3], and I would like to give again some precisions about its role in scientific creativity.

1 Deduction, induction, abduction (retroduction)

Let us begin with a quotation by Peirce:

> There are in science three fundamentally different kinds of reasoning. Deduction (called by Aristotle συναγωγή or ἀναγωγή), induction (Aristotle's and Plato's ἐπαγωγή), Retroduction (Aristotle's ἀπαγωγή) but misunderstood because of corrupt text, and as misunderstood usually translated abduction. Besides these three, Analogy (Aristotle's παραγωγή) combines the characters of Induction and Retroduction. (Peirce, CP I, 65)

A simple example illustrates these three inference processes: Induction (I) is generalized by a law (*"All crows are black"*) based on a correlation between observed facts (*"The crows that have been observed are all black"*)

---

[1] [Peirce 1965] quoted by "CP" (for *Collected Papers*) in this article.
[2] [Polya 1965/1989: 106].
[3] [Desclés 1996]; [Desclés 2000].

from a sample considered representative and large enough to be significant; Deduction (II) leads to a true statement (*This bird is black*) from two premises declared true (*"All crows are black"* and *"This bird is a crow"*); Abduction (III) (also called "retroduction" or "hypothesis" by Peirce) leads to the formulation of a plausible hypothesis (*"It is plausible that this bird is a crow"*) based on the attested fact (*"This crow-sized bird is black"*) and common knowledge (*"It is well known that crows are black"*).

> Retroduction is the provisional adoption of a hypothesis, because every possible consequence of it is capable of experimental verification, so that the preserving application of the same method may be expected to reveal its disagreements with facts, if it does so disagree. (Peirce, CP I, 68)

> Abduction is the process of forming an explanatory hypothesis. It is the only logical operation which introduces any new idea; for induction does nothing but determine a value, and deduction merely evolves the necessary consequences of a pure hypothesis. (...) Deduction proves that something must be; Induction shows that something is actually operative; Abduction merely suggests that something may be. (Peirce, CP V, 171)

The different reasonings are compared with different inference schemes as follows:

Induction (I)	*Deduction (II)*	*Abduction (III)*
$a_1, \ldots, a_i, \ldots, a_n$	$P(a)$	$Q(a)$
$P(a_i) \,\&\, Q(a_i) \;\; (1 \leq i \leq n)$	$(\forall x)[P(x) \Rightarrow Q(x)]$	$(\forall x)[P(x) \Rightarrow Q(x)]$
$(\forall x)[P(x) \Rightarrow Q(x)]$	$Q(a)$	is-plausible $(P(a))$
Induction shows that something is *actually operative*.	Deduction proves that something *must be*.	Abduction merely suggests that something *may be*.

Abduction is tantamount to imagining a plausible hypothesis intended to explain, with the help of an inferential process, certain facts, some may seem rather unexpected and *a priori* surprising. Inference by Abduction (III) is completely different with an inference by Induction (I) and it is not an inference by Deduction (II).

> Presumption [abduction] is the only kind of reasoning which supplies new ideas, the only kind which is, in this sense, synthetic. Induction is justified as a method which must in a long run lead up to the truth, and that, by a gradual modification of the actual conclusion. There is no such warrant for presumption. The hypothesis which it problematically concludes is frequently utterly wrong itself, and even method need not ever lead to the truth. (...) Its only justification is that its method is the only way in which there can be any hope of attaining a rational explanation. (Peirce, CP II, 777).

Inductive inference constructs *a general law* (i.e., an implication $[p \Rightarrow q]$ between two propositions) from *a set of correlations between different occurrences 'p_i'* of 'p' and different occurrences 'q_i' of 'q'. Following statistical considerations on correlations, the general law can take the following *probabilized form* $[p \Rightarrow$ is probable $(q)]$, which allows to deduce the probability of a conclusion from a fact-finding. Deductive inference constructs a consequence 'q' of a general law $[p \Rightarrow q]$ when a hypothesis 'p' is considered as a true proposition. Adductive inference constructs a plausible hypothesis from *a general law* $[p \Rightarrow q]$ and *a true proposition* 'q' (e.g., an observed fact), and, in this case, 'q' is considered a consequence of the 'p' hypothesis. With abductive inference the proposition 'p' can be false when the premises 'q' and $[p \Rightarrow q]$ are true. In an abductive process, the proposition 'p' *is only a plausible explanation* of the fact 'q'; the explanation of the observed proposition (a statement) must be found; in this case, the proposition 'q' functions as a clue in favour of the plausibility of the hypothesis 'p'.

Remark: The inference scheme of abduction is very different from the inference scheme of deduction by *modus tollens*:

Deduction (*modus ponens*)	Deduction (*modus tollens*)	Abduction
p	$\neg(q)$ (negation of q)	q
$[p \Rightarrow q]$	$[p \Rightarrow q] = [\neg(q) \Rightarrow \neg(p)]$	$[p \Rightarrow q]$
q	$\neg(p)$ (negation of p)	is-plausible (p)

2 Abduction is a cognitive inference process

The process of inference by abduction that proposes a plausible hypothesis about the occurrence of an observed fact is a cognitive process, perhaps specific to human cognition. It is used in everyday life, for example from the observation "*Hey, the road is wet*" (proposition 'q'), we can infer, by an abduction, that "*It rained*", hence a statements like "*So, it rained*", that is to say the enunciation of a plausible proposition 'is plausible (p)', constructed from the general law "*When it rains, the road becomes wet*", which is a matter of common knowledge. However, other explanations can replace this plausible hypothesis, for example "*The municipal sprinkler passed by there a short time ago*" linked to common knowledge "*If the municipal sprinkler passes the road becomes wet*".

Let us present an example given by G. Polya[4]. As the three ships sailing West have not seen land (China or India) appear on the horizon as indicated in Christopher Columbus' plans, the crew was planning to revolt; however, some of its members noticed the presence of birds around the boats; this

---
[4] [Polya, 1958/2008: 181]; [Polya, 1965/1989: 104–108].

observation triggers abductive reasoning based on knowledge of sailors *"birds fly around the boats on land"* ($[p \Rightarrow q]$); since they have seen more and more frequent flights of birds around the ships ('q'), it was very plausible that one was near land ('is-plausible (p)'), they waited before beginning a revolt; indeed, the sailor on the lookout soon shouted *"Earth!, Earth on the horizon, in front of us!"*. Thus, Christopher Columbus and his three ships were able to land a large island (the island of San Salvador) off the coast of this New World—not on the mainland of China or India as they believed—that will become America.

2.1 "Evidentiality" (or "mediativity") expressed by natural languages

The cognitive process of abduction reasoning is generally expressed by contextualized linguistic expressions. Many languages have grammatical systems to explicitly indicate by means of utterances specifying inferences by abduction; in these languages, the grammatical systems contain explicit grammatical morphemes grouped together under grammatical label of "evidentiality" (or "mediativity")[5]. The natural languages as Tuyuca, Tariana, Quechua, Kashaya[6] are examples of natural languages having an evidential system with more than one inferential morpheme depending on the type of inference; they use grammatical markers to express the enunciation of a plausible hypothesis from an abductive inference; the grammatical markers indicate to the co-enunciator that the enunciator has certain clues in favour of the plausibility expressed by the utterance; other natural languages do not express directly "evidentiality" by a system of specific grammatical markers but these languages can perfectly express this semantic notion. The clues in favour of a plausible hypothesis are not expressed in the enunciation of a plausible hypothesis but they can be specified when the reason for this plausible hypothesis is demanded; for instance:

– *Hey, it has rained.*

– *Why do you say that?*

– *Look! The road is wet* [it is a clue in favour of a plausible hypothesis].

Let us take the example of Panare, a Caribbean language of Venezuela with morphological mechanisms whereby speakers must specify whether the fact they are presenting has been personally verified, or whether it is a hypothesis based on observed clues and therefore simply plausible:[7]

---

[5][Guentchéva 1996]; [Guentchéva & Landburu 2007]; [Desclés & Guentchéva 2018, 2024].

[6][Barnes 1984] for Tuyuca; [Aikhenvald 2003] for Tariana; [Faller 2002] for Quechua; [Oswald 1986] for Kashaya.

[7][Mattéi-Müller 2007].

(1) *a-të-se*   *mën*   *kanawa* *Ehkara* *pana*
    Intr-go-PST:Imm   Cop:Inan car   Caicara   DIR
    'The car just left for Caicara.'

(2) *n-ti-yah*   *kën*
    3-go-PST:Rec   3Sg:An:NonVis
    'It has left.' [The speaker saw it go]

(3) *yu-të-hpë*   *mën*   *kën*
    3Intr-go-PERF:Infer Cop. 3Sg:An: NonVis
    'It has left.' [herefore, it must have left]
    [Description by the author: The speaker notes that the person's
    hammock is no longer there and infers that the person has left].

Thus, utterances (1) and (2) are distinct from (3). In (1), the verb form bears the suffix '*-se*', it indicates a declarative sentence referring to an empirically observed fact; (2) denotes the state resulting from the same observed fact; in (3), the speaker neither verbalizes the resultant state as in (2), nor the occurrence of a recent past event as in (1); in (3), relying on clues (for example the person's hammock which is no longer there) and shared knowledge (when you leave a place, you take your hammock with you), the enunciator expresses a hypothesis, deemed highly plausible, based on clues that the person has left. The grammatical marker '*-hpë*' is an evidential marker that expresses the result of an abductive inference. In different languages (as Albanian, Bulgarian, Farsi, Georgian, ...), the perfect has given rise to a series of perfect-like forms which can express abductive inference based on clues; this grammatical form is used by detectives to elucidate a crime by an abductive reasoning based on a set of observable clues (broken window, traces of blood or other indications), the most plausible hypothesis can be confirmed or infirmed by the discovery of new clues, as in Bulgarian (a southern Slavic language):

(4) *Kradecăt* *e*   *vljazal*   *v*   *kuxnjata*   *prez*
    thief.Art   be.PRES   enter.PAP.Pf   in   kitchen-ART   through
                                                                prozoreca
                                                                window.Art
    'The thief has entered the kitchen through the window.'

One finds the same type of examples in the Nakh-Daghestanian languages, such as Agul, where judging from chips and other visible clues (scratches, ...) the speaker verbalizes a hypothesis to explain the observed facts.

The fact that natural languages express evidential (or mediative) statements, through explicit markers (sometimes grammaticalized in some languages), to indicate that the statement is based on abductive reasoning (with the recognition of clues and relationships established between plausible hypotheses and clues), leads us to think that the process of reasoning by abduction is a cognitive capacity, probably specific to humans, linguistic expressions being the observable traces of this capacity.

2.2 Abduction in everyday life

We are all faced with stating plausible hypotheses that can be explained, if we have to justify ourselves, by referring to clues:

- *Hey, a wild boar has passed by.*
- *Why?*
- *Well, look at these traces; they are the hoof traces of a wild boar.*

The hypothesis put forward as plausible can still be contested, or even completely refuted:

- *The boss has not arrived yet.*
- *Why?*
- *Look at his car. It is not in the parking lot.*
- *The boss's car is broken. Yesterday, he had to go back home by taxi.*
- *Thus, the boss, who usually arrives quite early at the office, would probably be already there.*

In the enunciation by an enunciator, called 'EGO', of a plausible (mediative or evidential) hypothesis, we have four steps:

1°) Observation of an observed fact 'C' (sometimes may be surprising);

2°) This observation triggers the search for a link between this fact 'C' and another fact 'H' which is likely to be an explanation of 'C';

3°) A reasoning by an adductive inference: '$C \& [H \Rightarrow C] \vdash$ is-plausible (H)';

4°) Enunciation of the plausible hypothesis: 'EGO-DIT (is-plausible (H))'.

The existence of certain historical figures (Napoleon, Jesus Christ, ...) is accepted as a plausible hypothesis, which can sometimes become hardly questionable since it derives its justification from more or less strong clues: Napoleon left material traces (his bicorn, a coat, letters, stories about his life and his actions, ...):

> Numberless documents and monuments refer to a conqueror called Napoleon Bonaparte. Though we have not seen the man, yet we cannot explain what we have seen, namely, all these documents and monuments, without supposing that he really existed (Peirce, CP, 2.625).

Some historians have questioned the existence of Jesus Christ because they have not found enough irrefutable clues, others, have been convinced in this existence based on strong clues (various converging narratives, indirect testimonies, consequences of this existence, ...). For instance the historian Jean-Christophe Petitfils considers, along with other historians, that the Healthy Shroud of Turin conjures up relevant facts in favour of the existence of Jesus Christ, while the philosopher Michel Onfray defends the idea that Jesus Christ has never existed, being a simple construction of the mind; this viewpoint is criticized by various historians[8]. This controversy shows that, relying on the same set of clues, several plausible hypotheses can be considered and discussed without necessarily leading to acceptance or rejection of "the best hypothesis". George Polya[9] noted that two people, confronted with the same argument and applying the same plausible inferences, may honestly find themselves in disagreement.

The paleontologist is led to state the plausible proposition: *"The sea was to cover these places in very ancient times"*, following the discovery of fossils in the form of fish buried in the earth at the top of a hill, appealing to the general law: *"Fish live inside the seas"*. Peirce writes:

> Fossils are found; say, remains like those of fishes, but far in the interior of the country. To explain the phenomenon, we suppose the sea once washed over this land. (Peirce, CP, 2.625)

2.3 Abduction and the reasoning by a detective, by a physician or by legal experts

Sherlock Holmes' plausible hypotheses are, in fact, abductive inferences as perfectly established by the semioticians Umberto Eco and Thomas A. Sebeok.[10] If we analyze Sherlock Holmes' method, we find out that what the detective (alongside with the author Conan Doyle) means when talking about Deduction and Observation, is, in fact, inference similar to Peirce's abduction. It is interesting to note that the above semioticians have compared the detective's reasoning to a physician's reasoning who seeks to observe the presence of certain symptoms to identify, as a result of abductive reasoning, a disease that would be the cause of these clues. In Umberto Eco's *Le Roman de la Rose*, Guillaume de Baskerville, in explaining the method followed, begins by discarding the idea of deduction as well as that of induction, and goes on to describe what Peirce calls abduction. In the domain of legal expertise, George Polya[11] gives excellent examples of heuristic inferences by abduction.

---

[8] [Petitfils 2022]; [Onfray 2023].
[9] [Polya 1958/2008: 234].
[10] [Levesque 2016].
[11] [Polya 1958/2008: 171–181].

3 Fundamental properties of abduction

Some of observed facts may be "surprising" and inexplicable outside the forwarded hypothesis. The surprising facts fall under the name of *serendipity*[12]. In order to be explained, the phenomena called serendipity often lead to triggering reasoning by abduction in order to be explained. However, in an adductive inference, the findings that trigger this inference are not necessarily "surprising", they may be perfectly "normal". The explanatory hypothesis, however, remains simply plausible and may be opposed by other equally plausible hypotheses.

A plausible hypothesis can be rejected, it is the epistemological force of abduction; the hypothesis *that explains one or more facts is not a truth*, it is always refutable in particular following the fact '$\neg(C)$' (negation of 'C') which contradicts what must be "normally" deduced from this hypothesis; by following this negative observation, the hypothesis must be rejected:

$$[H \Rightarrow C] \,\&\, \neg(C) \vdash \neg(H).$$

In some cases, the assumption may be modified and adjusted to take into account this negative fact.

A plausible hypothesis can be justified and reinforced by a bundle of concordant clues. The abductive scheme of inference becomes:

$$[H \Rightarrow (C_1 \,\&\, C_2 \,\&\, \ldots \,\&\, C_n)] \,\&\, (C_1 \,\&\, C_2 \,\&\, \ldots \,\&\, C_n) \vdash \text{is-plausible } (H).$$

The bundle of observed clues '$C_1 \,\&\, C_2 \,\&\, \ldots \,\&\, C_n$' reinforces the plausibility of the explanatory hypothesis. For instance, the observations that the orbits of different planets are ellipses reinforce the plausibility of the Copernicus' heliocentric system.

Several plausible hypotheses 'H_1' and 'H_2' can often co-exist; both 'H_1' and 'H_2' hypotheses can explain the same facts:

$$C \,\&\, [H_1 \Rightarrow C] \vdash \text{is-plausible } (H_1),$$
$$C \,\&\, [H_2 \Rightarrow C] \vdash \text{is-plausible } (H_2).$$

As long as one does not discover facts that allow rejecting one of the hypotheses, both hypotheses must be *a priori* accepted as plausible. Thus, for a lot of philosophers, theologians, astrologers, during several years, the geocentric system co-exited with the heliocentric system; the two systems explained the same observations (but by different ways). We have seen above that the examination of the real existence of Jesus Christ leads to two plausible hypotheses that prove incompatible when taking into account the same clues provided by historical documents. Polya[13] evokes a discussion

---

[12][Andel & Bourcier 2009].
[13][Polya 1958 / 2008: 233–234].

about the value of a plausible hypothesis in mathematics and he notes that two people, confronted with the same argument, applying the same schemes of plausible inference, may in all honesty find themselves in a disagreement.

4 What abduction is not

Plausible hypothesis built by abduction is not (necessarily) "the best hypothesis"; some philosophers of science defend this feature of abduction[14]. For us, this is not admissible since several plausible hypotheses may explain the same observed facts but other considerations must also be forwarded to prefer a hypothesis and to reject another. When several plausible hypotheses are in competition, some researchers might prefer one hypothesis for simplicity reasons, ability to explain many other facts, and even aesthetics to satisfy the Ockam's razor. For instance, Copernicus' heliocentric hypothesis is much simpler than Tycho-Brahe's geocentric system, which must use many epicycles to account for the many observations, and the Copernican system, defended by Galileo, enabled to define laws that took mathematical forms and later lead to Newtonian laws.

It is essential not to confuse on one hand, *the enunciation of a probable consequence of a fact* and on the other hand, *the enunciation of the plausibility of a hypothesis from an observed fact* interpreted as a clue. Indeed, the contexts of these two enunciations are often entirely different. Let's take two different contexts. Context I: this morning, people discover corpses on the beach (an observed fact 'q') and it is shared knowledge that when there is a shipwreck, corpses always wash up on the beach; thus one person can say: "*Therefore, there must have been a shipwreck the other night*"; 'is-plausible (p)' is inferred by reasoning by abduction from the clue 'q'. Context II: there was a shipwreck during the night (it is a fact 'p') and this morning, one people can say: "*There will probably be corpses on the beach*"; 'is-probable (q)' is a consequence of 'p' because it is common knowledge that when there is a shipwreck in the vicinity, the corpses of the castaways often wash up on the beach (the implication $[p \Rightarrow$ probability $(q)]$ is common knowledge). To give an example of this difference let us take Pomo language where Robert Oswalt isolates two suffixes '*-qă*' and '*-bi*' in his grammar of Kashaya[15]; he distinguishes (5a) from (5b) with two different interpretations, but, unfortunately, with a same translation in English:

(5) a. *sinam**q**^{h}*
    drown-INFER.I
    'He must have drowned'

b. *sinamq$^?$**biw***
    drown: INFER II: ABS
    'He must have drowned'

---

[14] [Walton 2004] for instance.
[15] [Oswalt 1961: 243].

of the two inference schemes relative to context I (with a plausible hypothesis) and context II (with a probable consequence):

Context I: q (constat) & $[p \Rightarrow q] \vdash$ is-plausible (p);

Context II: p (constat) & $[p \Rightarrow$ is-probable $(q)] \vdash$ is-probable (q).

in Kashaya, when a person enters a house and detects the smell of baked bread, he could say either (6) or (7):

(6) cuhni· mu$^?$'ta-**q**^{h}
    bread cook-INFER.I
    'Bread has been cooked'

(7) cuhni· mu$^?$'ta mihšew
    bread cook smell
    'It smells like cooked bread'

In sentence (6), the smell is a clue, hence the inference of a highly plausible hypothesis: "bread has been cooked". In contrast, in (7), there is no inference and the verb is used simply to declare a direct olfactory perception.

5 How to check the accuracy of an abduction?

The formulation of a plausible hypothesis from a reasoning by abduction leads quite naturally reinforcing the plausibility of the hypothesis by resorting to statistical correlations between the hypothesis and the occurrences of observed cases, so as to be able to pose the general law: $[H \Rightarrow C]$. Here, the induction which concludes with the formulation of a general law is guided by the hypothesis 'H' that should be confirmed or rejected when the number of proven correlations (the sample) is considered too low. When the inductive test is positive, the reasoning by abduction takes the form: C & $[H \Rightarrow C] \vdash$ is-plausible (H); in this case, the plausible hypothesis 'H' can be accepted (at least provisionally) as an acceptable scientific hypothesis (therefore assumed to be true) which becomes an explanation of the observed case 'C'.

> The induction adds nothing. At the very most it corrects the value of a ratio or slightly modifies a hypothesis in a way which had already been contemplated as possible. (Peirce, CP VII, 217)
>
> For abduction commits us to nothing. It merely causes a hypothesis to be set down upon our docket of cases to be tried. (Peirce, CP V, 602)
>
> [...] the entire meaning of a hypothesis lies in its conditional experiential predictions: if all its predictions are true, the hypothesis is wholly true. (Peirce, CP VII, 203)

Abduction can sometimes lead to dead ends. The study of anagrams by Ferdinand de Saussure is a very good example; in the latter part of his life, Saussure became passionate about the study of anagrams, trying to discover hidden hypotheses, a kind of "occult traditions" in Greek and Latin poetry[16]; he first proposed hypotheses and then complicated them by other hypotheses. The inductive verifications led him to find out that there were practically no restrictions (no laws) all the constructions examined could support the hypothesis put forward on the anagrams. He gave up this research.

6 Examples of discoveries from reasoning by abduction

Let us quote different examples of formulations of plausible hypotheses in different domains of sciences (natural and human sciences).

6.1 Plausible hypotheses in mathematics and astronomy

In the field of mathematics, the reasoning by abduction (under the name of "heuristic reasoning") is the discovery and formulation of a new plausible proposition that must then be demonstrated to make a mathematical truth. For example, the Fermat's conjecture (for $n > 2$: $[a^n + b^n = c^n]$ is impossible), is a plausible hypothesis whose justification can be based on a large number of consequences demonstrated as true; these demonstrated consequences confirm the plausibility of the conjecture but they are not a proof; despite many efforts, Fermat's conjecture has not been demonstrated for three centuries, but, finally, Andrew Wiles, in 1993, has given a proof, which after many verifications, has been accepted by the community of mathematicians[17]. At present, the Riemann's conjecture, which aims to shed light on the infinite distribution of prime numbers, has not been proved yet.

In the field of astronomy, Polya[18] traces Kepler's different hypotheses and rejects them. Kepler seeks to discover the cause or some reason for the number of planets, their distance from the sun and the periods of their revolutions; he imagines 11 concentric surfaces, 6 spheres alternating with 5 regular polyhedra. The first surface is external to the others, it is a sphere and each surface is encompassed by the previous one; to each sphere is associated a planet, the radius of the sphere gives the (average) distance of the planet to the sun. Each polyhedre is inscribed in the previous sphere and is circumscribed to the next sphere. Kepler compares the plausible hypothesis with observations. The expected agreements are good in some cases and very bad in others. Kepler must therefore modify his initial hypothesis while remaining faithful to his preconceived idea: the sphere is the "perfect figure" and the five regular polyhedra, the figures of Plato,

---

[16][Starobinnski, 1971]; [Fadda 2018: 25–28].
[17][Hellegouarch, 1997]; [Singh 1998].
[18][Polya 1958/2008: 137–140].

are the "noblest figures". It therefore seems "natural" to him that the sun and the planets are in a certain way linked to the figures of Euclid. Polya notes that the confidence we place in a hypothesis depends on the cultural environment and the scientific atmosphere of a period; he also emphasizes Galileo's intellectual courage and his independence of mind facing prejudice of his time while Kepler, a contemporary of Galileo, was influenced by the mysticism and the prejudices of his time.

6.2 Non-observable plausible hypotheses in relation to observables in physics

We can quote several examples of plausible hypotheses proposed by researchers without direct observable correspondents. These assumptions can then slightly to be adjusted and finally accepted, for example from the results of new observations, or they may be heavily modified and sometimes rejected. Jean Perrin formulated the atomic hypothesis of atoms (with protons with electrons around). Criticized in the beginning, this hypothesis has finally been accepted by the entire community before being seriously refined by contemporary physics. Albert Einstein, who defended, for *a priori* ideological reasons, the hypothesis of a homogeneous isotropic Universe, preferred to modify the equations of General Relativity by introducing a "cosmological constant" that preserved the stability of the Universe. Faced with a large number of empirical results, Einstein will recognize his error (*"the greatest mistake of my life"*) and return to this cosmological constant by accepting the hypothesis of a dynamic Universe that expands (or contracts).

Drawing certain consequences from Einstein's General Relativity, the physicist Georges Lemaître formulated in 1927, after Alexander Friedmann (1922), the hypothesis of the "primitive atom", which assumed a temporal beginning of the Universe, that is to say the hypothesis of the "Big Bang" highlighted by Edwin Huggle in 1929. This hypothesis was opposed to the idea of a stable and eternal Universe, commonly accepted at this time. The discovery in 1965 by Arno Penzias and Robert Wilson of an "echo" of a cosmic microwave background confirmed the dynamic cosmological scenario of a rapid expansion of the Universe from an extremely dense and extremely hot state. This Big Bang hypothesis has given rise to many philosophical interpretations. The Big Bang hypothesis is now accepted as plausible but not in the form of "a primitive atom", extremely dense and hot, which would have exploded and separated on one side a nothingness and on the other hand, a world where time and space took shape. The plausible hypothesis of the Big Bang leads to the formulation of many scientific and philosophical problems that do not yet find real answers.

In the field of quantum physics, confronted with the phenomena of interactions at the atomic and subatomic level, it is necessary to formulate many plausible hypotheses at the source of mathematical calculations that account for experimental results but as Richard Feynman[19] says, we are not really sure that we have yet really "understood" the world of quantas.

6.3 Two examples of a discovery of a plausible hypothesis in linguistics

The two following examples, borrowed from linguistics, clearly show that there are scientific approaches in the human sciences, as in the natural sciences, which lead to very solid results capable of garnering the support of specialists in the discipline.

Jean-François Champollion (1822) has been able to justify his plausible hypothesis following a succession of more or less refined hypotheses and the rejection of false hypotheses. Having had access to new documents (the Huyot documents), Champollion decided to apply the writing system used for the names of Greek kings to the names of the rulers of the high Egyptian Empire. By analyzing new names, he formulated the hypothesis of the triple writing system of Egyptian hieroglyphics, which are, for some, phonetic inscriptions, for others, ideographic and also symbolic inscriptions[20]. Before the formulation of this fruitful hypothesis, for many years, Champollion, following Sylvester de Sacy, defended the exclusively ideographic nature of Egyptian hieroglyphics, and he persisted in believing that this was a self-evident fact until the evidence of the facts led him to recognize the phonetic value of a group of hieroglyphics constituting the inscriptions that decorated Egyptian monuments of different periods. The adductive approach undertaken by Champollion allowed to obtain solid results that the method of its competitor Thomas Young could not achieve.

The young Ferdinand de Saussure (1879) formulates an plausible hypothesis about the proto-Indo-European language. There are three major periods in Ferdinand de Saussure's work: (i) the period of youth with the publication of the *Mémoire* (called *Le Système primitif des Voyelles dans les langues Indo-européennes*) presented in Leipzig in 1879, which made him noticed among the linguists of this time; (ii) the period of the courses professed at the University of Geneva with the publication of the famous *Cours de Linguistique générale* (written by three of the course's auditors), which earned Saussure to be considered one of the great founders of structural linguistics and general linguistics; (iii) the period of research on the anagrams (mentioned above). It is the first period, that of the *Mémoire*, which interests us here. By examining the systems of vowel alternations in several known languages (Greek, Sanskrit, Latin, Germanic languages, ...) and based on general rules of diachronic changes formulated by different linguists, Saussure, only 21 years old, formulates this hypothesis: "A cer-

---

[19][Feynman 1965].
[20][Lacouture 1988]; [Desclés 2000: 97–99].

tain phoneme, not attested in the languages studied (of the Indo-European family), exists in the proto-Indo-European language; this phoneme would make it possible to explain all the phenomena attested in all the studied languages of this family of languages". This phoneme is a laryngal that Saussure calls 'coefficient sounding'; this reconstructed phoneme is absent in all the languages hitherto observed but its plausible existence made it possible to explain some embarrassing phenomena. The plausibility of this hypothesis is based on laws of phonetic change formulated, at this time, by the works about the comparison of studied indo-european languages. The Saussure's hypothesis makes it possible to link this phoneme, not empirically directly observed, to a certain number of phonemes that are attested in different studied indo-European languages. It was only in 1927, after the deciphering of the Hittite language by F. Kurilowicz, that was actually observed a phoneme which Saussure's reasoning had put in place in the form of a plausible hypothesis about a proto-Indo-European language[21].

6.4 Semantic representations related to linguistic expressions

Nowadays, cognitive linguistics uses semantic-cognitive representations—unobservable—that have grammatical markers (tenses, aspects, various modalities, determination) as observable traces in the semiotic systems of natural languages. The lexical units of verbs and prepositions have meanings that are described precisely by more abstract cognitive representations, obtained by composing cognitive "primitives" closely related to perception, action and agents with more or less intentional aims, in nested and entangled relationships. The linguist Sebastian Shaumyan[22], taking up a distinction of a biological nature between genotype and the various phenotypes, undertook the description of the Grammar of a genotype language—not accessible to direct observation—with two levels of description: on the one hand, the linguist must describe the main invariant constructions of language activity; on the other hand, he must link these invariants to the different observed phenotype languages, semiotic systems organized by the specific rules of these natural languages. Let us take an example: by differentiating according to the order of grammatical and lexical units in standard sentences, some natural languages (Ancient Greek, Latin, Arabic, ...) grammaticalize explicitly certain constructions with mandatory morphological cases but not other languages (as English, French, ...). The two models, 'Applicative and Cognitive Grammar' (GAC) and 'Applicative, Cognitive and Enonciative Grammar' (GRACE)[23], develop Shaumyan's ideas, by linking plausible semantic-cognitive representations, not directly observable, to the observable semiotic forms of natural languages, by means of intermediary changes of

---

[21][Apresjan 1973: 98–101]; [Desclés 2000: 99–102].
[22][Shaumyan 1977, 1987].
[23][Desclés 1990]; [Desclés et al., 2016].

representations expressed by inferences formulated in the formalism of the Combinatory Logic of Curry[24], a logic of whatever operators intrinsically combined and transformed by abstract operators, called "combinators" (according to a general hypothesis of compiling between different levels of representations)[25].

7 Conclusions

In conclusion, let's listen again to Peirce:

> Abduction is the process of forming an explanatory hypothesis. It is the only logical operation which introduces any new idea; for induction does nothing but determine a value, and deduction merely evolves the necessary consequences of a pure hypothesis. Deduction proves that something must be; Induction shows that something is actually operative; Abduction merely suggests that something may be. (Peirce, CP V, 171)

> Presumption [abduction] is the only kind of reasoning which supplies new ideas, the only kind which is, in this sense, synthetic. Induction is justified as a method which must in a long run lead up to the truth, and that, by a gradual modification of the actual conclusion. There is no such warrant for presumption. The hypothesis which it problematically concludes is frequently utterly wrong itself, and even method need not ever lead to the truth. (...) Its only justification is that its method is the only way in which there can be any hope of attaining a rational explanation. (Peirce, CP II, 777)

From what we have just recalled in this article, it becomes clear that the formulation of a creative hypothesis does not emerge from big data. The creative hypothesis aiming to "explain" a problem or some questions that a researcher has been able to discover and to formulate this hypothesis in precise terms, is often adductive: 1°) he observes problematic facts that are not explained (sometimes surprising and going against common knowledge); 2°) to explain these facts, he formulates a new hypothesis 'H' which maintains relations of implication with these problematic facts, this is the important moment of explanatory creativity; 3°) he infers, by a reasoning by abduction, that the hypothesis 'H' is plausible; 4°) this hypothesis 'H' would thus explain (at least provisionally) the nature of the problem raised by finding the observed problematic facts. Thus, the researcher and his community (scientific, cultural, social community etc.) should seek to strengthen the plausibility of the hypothesis stated, by examining the consequences of the hypothesis or, sometimes, by accepting that this hypothesis, supported as only plausible and therefore fallible, must ultimately be rejected or, in

---

[24][Curry et al. 1958, 1972].
[25][Desclés 2004]; [Desclés et al. 2016].

some cases, must be entirely reformulated to fit more accurately to the consequences of the plausible hypothesis. The accumulation of data rarely leads to the formulation of a new hypothesis capable of explaining and understanding a certain number of problems that informed and attentive minds have been able to identify; on the other hand, the accumulated data are an adequate place where a plausible hypothesis can be confirmed or rejected.

References

[Aikhenvald 2003] Aikhenvald, Alexandra (2003). *A Grammar of Tariana, from Northwest Amazonian*, Cambridge: Cambridge University Press.

[Andel & Bourcier 2009] van Andel, Pek & Bourcier, Danièle (2009). *De la sérendipité dans la science, la technique, l'art et le droit*, ACT MEM.

[Apresjan 1973] Apresjan, Juri Derenikowitsch (1973). *Éléments sur les idées et les méthodes de la linguistique structurale contemporaine*, Paris: Dunod.

[Barnes 1984] Barnes, Janet (1984). *Evidentials in Tuyuca Verbs*, IJAL 50, pp. 255–271.

[Curry et al. 1958] Curry, Haskell & Feys, Robert (1958). *Combinatory Logic*, Vol. I, Amsterdam: North Holland Publishing.

[Curry et al. 1972] Curry, Haskell, Hindley, J. Roger, & Seldin, Jonathan P. (1972). *Combinatory Logic*, Vol. II, Amsterdam: North Holland Publishing.

[Desclés 1990] Desclés, Jean-Pierre (1990). *Langage applicatifs, langues naturelles et cognition*, Paris: Hermès.

[Desclés, 1996] Desclés, Jean-Pierre (1996). L'abduction, procédé d'explication en linguistique, *Modèles linguistiques*, Tome XVII, Fascicule 2, pp. 33–62.

[Desclés 2000] Desclés, Jean-Pierre (2000). *Abduction and non-observability. Some Examples from Language and Cognitive Sciences*, in Evandro Agazzi and Massimo Pauri (eds) 2000. *The Reality of the Unobservable*, Dordrecht: Springer, pp. 87–112.

[Desclés 2004] Desclés, Jean-Pierre (2004). *Combinatory Logic, Language, and Cognitive Representations*, in Weingartner (ed.) 2004. *Alternative Logics. Do Sciences Need Them?* Berlin: Springer, pp. 115–148.

[Desclés et al. 2016] Desclés, Jean-Pierre, Guibert, Gaëll, & Sauzay, Benoît (2016). *Logique combinatoire et lambda-calcul: des logiques d'opérateurs* (volume I); *Calculs de significations par une logique d'opérateurs* (volume II), Toulouse: Cépadues.

[Desclés & Guentchéva 2018] Desclés, Jean-Pierre & Guentchéva, Zlatka (2018). *Inference processes expressed by languages: Deduction of a probable consequent vs. abduction*, in Arigue & Rocq-Miguette (2018). *Theorization and Representations in Linguistics*, Cambridge Scholars Publishing, pp. 241–265.

[Desclés & Guentchéva 2024] Desclés, Jean-Pierre & Guentchéva, Zlatka (2024). *Évidentialité, médiativité, modalité épistémique, une approche énonciative, Conférence au Congrès Mondial de Linguistique Française*, CMLF 2024, Lausanne, 4 July 2024.

[Fadda 2018] Fadda, Emanuele (2018). *Abduction et firstness: la réalité du possible et la possibilité du réel*, in Clot-Goudard, Remi, et al. (eds; 2018). *Abduction, Recherches sur la philosophie et le langage*.

[Faller 2002] Faller, Martina (2002). *Semantics and Pragmatics of Evidential in Cuzco Quechua*, PhD Thesis, Stanford University.

[Feynman 1965] Feynman, Richard P. (1965). *The Feynman Lectures on Physics*, California Institute of Technology: Addison-Wesley Publishing Company, inc Reading, Massachussetts.

[Guentchéva 1996] Guentchéva, Zlatka (ed.; 1996). *L'énonciation médiatisée*, Paris-Louvain: Peeters.

[Guentchéva & Landaburu 2007] Guentchéva, Zlatka & Landaburu, Jon (eds.; 2007). *L'Enonciation médiatisée II: Le traitement épistémologique de l'information: Illustrations amérindiennes et caucasiennes*. Paris: Peeters.

[Hellegouarch, 1997] Hellegouarch, Yves (1997). *Invitation aux mathématiques de Fermat-Wiles*, Paris: Masson.

[Lacouture 1988] Lacouture, Jean (1988). *Champollion, une vie de lumière*, Edition Grasset et Fasquelle.

[Levesque 2016] Levesque, Simon (2016). Le Signe des Trois. Dupin, Holmes, Peirce d'Umberto Eco and Thomas Sebeok, *Cygne noir* 4.

[Mattéi-Müller 2007] Mattéi-Müller, Marie-Claude (2007). Voir et savoir en panaré (langue caribe du Venezuela), in [Guentchéva & Landaburu 2007: 153–169].

[Onfray 2023] Onfray, Michel (2023). *Théorie de Jésus*, Paris: Le Bouquin.

[Oswald 1961] Oswald, Robert (1961). *A Kasheya grammar (Southwestern Pomo)*, PhD Thesis, University of California at Berkeley.

[Oswald 1986] Oswald, Robert (1986). The evidential system of Kashaya, in Chafe & Nichols (eds; 1986). *Evidentiality. The Linguistic Coding of Epistemology*, Norwood N.J.: Ablex, pp. 29–45.

[Peirce 1965] Peirce, Charles Sanders (1965). *Collected Papers of Charles Sander Peirce*, Vol. I–VI, edited by Charles Hartshorne and Paul Weiss, Cambridge, Massachusetts: The Belknap Press of Harvard University Press.

[Petitfils 2022] Petitfils, Jean-Christophe (2022). *Le Saint Suaire de Turin Témoin de la Passion de Jésus Christ*, Paris: Editions Tallandier.

[Polya 1958/2008] Polya, Georges (1958). *Les mathématiques et le raisonnement plausible*, Paris: Gauthier-Villars; new edition: 2008. Paris: éditions Jacques Gabay.

[Polya 1965/1989] Polya, Georges (1965). *Comment poser et résoudre un problème, Mathématiques – Physique – Jeux – Philosophie*, Paris: Dunod; new edition 1989. Paris: éditions Jacques Gabay.

[Shaumyan 1977] Shaumyan, Sebastian K. (1977). *Applicational Grammar as a Semantic Theory of Natural Language*, Chicago: Chicago University Press.

[Shaumyan 1987] Shaumyan, Sebastian K. (1977). *A Semiotic Theory of Language*, Bloomington and Indianapolis: Indiana University Press.

[Singh 1998] Singh, Simon (1998). *Le dernier théorème de Fermat*, Paris: JC Lattès.

[Starobinnski 1971] Starobinnski, Jean (1971). *Les mots sous les mots. Les anagrammes de Ferdinand de Saussure*, Paris: Gallimard.

[Walton 2004] Walton, Douglas N. (2004). *Adductive Reasoning*, Tuscaloosa: The University of Alabama Press.

Continuity and discontinuity in theory change

Dennis Dieks

History and Philosophy of Science, Freudenthal Institute, Universiteit Utrecht, Buys Ballotgebouw (BBG), Princetonplein 5, 3584 CC Utrecht, The Netherlands
E-mail: d.dieks@uu.nl, ORCID: 0000-0003-4755-3430

Abstract. According to the currently most popular version of scientific realism, the growing success of science is explained by the way successive scientific theories preserve what was true in older theories while replacing theoretical parts that have been proven false. According to this accumulative realism, it is true that scientific changes can introduce radically new ideas. But on closer inspection, there is also considerable preservation of fundamental truths or approximate truths. This view justifies the idea that successive theories get closer and closer to the truth by eliminating errors and adding to what has already been shown to be correct. Here we present an alternative to this accumulative view of scientific progress. We point out that successful parts of older theories are usually not adopted into new theoretical frameworks, but rather *emerge* as approximations with limited applicability. These emerging patterns are derived within a new theoretical framework that may be completely different from that of the old theory. Thus, the changes resulting from theory replacement are often more drastic than expected based on realistic intuitions. This argument casts doubt on the idea that science develops cumulatively, by accumulating more and more pieces of truth.

1 Introduction

According to scientific realism science aims at representing the world as it really is, both concerning what is observable and what is unobservable. It is an epistemically optimistic doctrine, not only saying that science has the aim of finding out the truth about the physical world but also claiming that science possesses the means to achieve this aim. Our present scientific theories, which have developed since the scientific revolution and have reached impressive predictive success can accordingly be trusted to already contain a good deal of theoretical truth. Indeed, a typical realist argument runs, it would be miraculous if science had the predictive and explanatory success it actually has, if it did not latch on to what is really going on in nature, also at the level of the unobservable. This is the so-called "no-miracles argument" for scientific realism, according to which doubting that science describes the actual mechanisms responsible for observable phenomena would amount to attributing the empirical success of science to the miraculous coincidence of finding incorrect theories that happen to yield correct predictions.

However, there is an obvious counterargument. Time and again during the history of modern science, empirically successful and seemingly unassailable theories have eventually proven to be inadequate. For example, Newton's mechanics was once considered the epitome of what could be achieved in natural science, and it seemed absurd to doubt its principled truth. It was even widely regarded as an ideal to deduce the fundamentals of other disciplines from Newtonian principles, in order to secure their truth. Nevertheless, this monument of successful physics began to falter at the end of the 19th century and has now long since been replaced by the radically different quantum mechanics. Generalizing from such cases, it appears likely that our present theories will eventually prove inadequate as well; in other words, we have to assume that they are false. This would imply that their undeniable empirical successes do not provide convincing evidence for the truth of their assumptions about underlying processes and entities. This is the so-called "pessimistic meta-induction" (Laudan 1981).

The realist camp, however, does not yield so easily. According to realists, it must be admitted that in the process of replacing a theory, some ideas about the nature of the physical world are usually overturned and some theoretical axioms are rejected; and that in this sense the replaced theory as a whole was false. But this does not mean, realists claim, that the replaced theory contained no truth. Realists claim that a detailed look at the history of science shows that not everything is thrown overboard during theoretical changes. On the contrary, some central elements of the old theory usually remain, perhaps in a refined form. Further, it should be expected that it is precisely these retained elements that were responsible for the predictive success of the old theory. Thus there is, after all, a continuous accumulation of truth. Faced with the pessimistic meta-induction, the realist needs only make a small concession, namely, that it is overly optimistic to believe in the truth of what a theory says *in toto*. But this does not change the fact that the success of a theory indicates that part of it is true or approximately true. One must be careful and selective and limit one's confidence to the approximate truth of those theoretical parts that were essential in producing successful predictions. These true parts are retained, which legitimizes the view that successive scientific theories get progressively closer to the truth.

This article critically examines this accumulative realist view according to which the history of science shows a continuity between successive theories that demonstrates the gradual refinement and extension of previously achieved partial truths. Certainly, we must admit that there is some kind of continuity between successive scientific theories: without it, new theories would not be able to reproduce the successes of their predecessors. However, we will argue that the continuity in question is typically the result of what is called 'emergence' in the philosophy of physics. The term 'emergence' refers to patterns and regularities that are unexpected on the basis of the fundamental laws of a theory, yet occur within a limited part of the theory's application domain; they are approximate and typically occur in coarse-grained quantities when calculated in limiting situations. In principle, it is always possible to show that such emergent patterns lack fundamentality, in the sense that the often drastically different fundamental laws of the theory still apply and can yield more accurate predictions and explanations.

2 Retention versus emergence

(Some of the material of the following sections is also covered in (Dieks 2023c), on which the present presentation improves.)

The realist response to the pessimistic meta-induction hinges on the notion that in periods of theory change theories may well undergo drastic changes, but that a number of features such as causal mechanisms, sets of equations, or selected axioms, are typically retained and incorporated into successor theories. This preservation of theory parts is taken to indicate that the superseded theories included a kernel of truth or approximate truth. The empirical success of the older theories can be explained by their true parts (Psillos 1994, 2009, 2022; Alai 2021); that empirical success was consequently anything but miraculous, even though the older theories were strictly spoken false. As science advances, incorrect aspects of theories are gradually removed while true components are retained, extending the set of uncovered truths and improving our understanding of the world.

A standard illustration of this realist response is the transition from Maxwell's 19th-century electromagnetic theory to Einstein's 1905 electrodynamics. Maxwell's theory aimed at explaining electromagnetic phenomena as manifestations of mechanical processes, vibrations, in a material medium, the "ether", that filled the entire space of the universe. But in 1905 Einstein published his special theory of relativity, in which the same electromagnetic phenomena were accounted for without invoking any ether-like mechanical substratum. This was a revolutionary change in ontology, hard to digest for many physicists and only gradually accepted by the scientific community. However, despite this major ontological upheaval, the mathematical equations interrelating charges, currents, fields, and forces remained the same in the new theory. And of course, it was these equations that had made the successful predictions of Maxwell's theory possible; the interpretation of electric and magnetic fields as vibrations in an underlying mechanical medium played no role in the mathematical derivations. In this historical example there clearly is a theoretical core part that was retained: the relations between electromagnetic quantities represented by Maxwell's equations were left untouched. The *theoretical structure* of Maxwell's theory, defined by the relations between quantities as specified by the Maxwell equations, may thus plausibly be viewed as representing a truth already present in 19th-century electrodynamics, and as such only to be expected to survive the Einsteinian revolution (Worrall 1989). By contrast, the false assumption that electromagnetic phenomena possess a mechanical character was rightly discarded, in agreement with the core idea of accumulative selective realism.

It should be noted, however, that the Maxwell-Einstein case is atypical: it hardly ever occurs in modern physics that portions of basic mathematical formalism remain completely intact when transitioning from one theory to another. In this respect, it is interesting to compare the following example, the transition from the 18th-century caloric theory of heat to modern thermodynamics.

The key idea of caloric theory is that heat behaves as a fluid. Heat is assumed to be a conserved substance, "caloric", consisting of very small particles that repel each other but are attracted by other matter. This theory achieved considerable empirical success (for instance, it provided elegant explanations for the expansion of materials when heated, for the fact that heat flows from hot to cold places and not from cold to hot, and for many other thermal phenomena). However, the caloric theory was completely rejected in the 19th century because its predictions failed in important cases (e.g., the production of heat by rubbing objects vigorously). According to its successor, thermodynamics, heat is not a material substance but rather a form of energy. Work, another form of energy, can be converted into heat so that heat cannot possibly be a conserved quantity. Despite this radical rejection of the core idea and ontology of caloric theory, defenders of accumulative scientific realism claim that elements of "caloric explanations" are still recognizable in explanations given by modern thermodynamics. For example, in some cases, when there is no conversion of work into heat, conservation of energy can play the same role as the earlier principle of conservation of caloric. Then again, in certain specific cases, caloric can be said to have had the same function as nitrogen in the 19th-century theory of heat; in certain other specific cases, it behaved much like modern oxygen. One might therefore argue that caloric theory was partially, approximately, and "locally" on the right track, specifying mechanisms in specific cases that bear a resemblance to what modern theory says in those same specific cases. In this way, the idea that elements of truth contained in caloric theory are preserved in successor theories may still be defended (Psillos 1994), despite the fact that the outlook of caloric theory is radically different from its modern counterparts. The case is certainly less clear than that of the Maxwell-Einstein transition however, and the claim that we are facing a case of truth retention here remains controversial (see, for example, Chang 2003, and the overview Psillos 2022, with references to criticisms contained therein; also Cordero 2011 for critical discussion of the Maxwell-Einstein case). Anyway, that successes of caloric theory can be reinterpreted by the modern theory of heat in locally structurally similar ways need not surprise us: modern theory should evidently be able to reproduce old successes, and since the mechanisms proposed by caloric theory closely follow directly observable regularities there is little reason to expect that newer theories would use structurally very different local explanations. An appeal to deeper truth seems unnecessary (see section 4).

A somewhat similar historical case may highlight implausible aspects of seeking truth in superseded theories at all costs. This example goes back to the beginnings of science. The germination of modern science is usually associated with the rejection of Aristotelianism: it is widely accepted that the Aristotelian physical world picture is fundamentally misguided and that the scientific revolution could only succeed when Aristotelian dogmas were left behind.

We will focus here on the relation between Aristotelian mechanics (Aristotle's theory of motion) and Newtonian, so-called classical, mechanics. One of the important differences between Newtonian and Aristotelian mechanics is that according to the former theory, material bodies on which no forces act persist in a state of uniform motion. Forces are therefore not needed to *maintain* motion; instead, they cause states of motion to *change*. Forces accelerate material bodies, according to the famous equation $F = m \cdot a$. By contrast, according to Aristotelian mechanics, a body will remain at rest unless a force compels it to move. Aristotle posits that forces produce a *velocity*, and instead of the Newtonian law of motion $F = m \cdot a$ there is the Aristotelian principle $v = \frac{F}{R}$, where v, F, and R denote the velocity of a moving body, the force exerted on it, and the resistance offered by the surrounding medium, respectively.

But even though Newton's mechanics describes the physical universe and its fundamental principles in a way that is completely incompatible with the Aristotelian view, one should expect some continuity between the two theories. Aristotle's mechanics could not have survived so long if there had been no empirical support. In fact, many everyday observations can easily be accommodated within the Aristotelian framework: objects around us do not begin to move of their own accord. We must exert a force to make them move and to maintain their motion. Empirical facts of this sort should obviously be explainable by classical mechanics as well. So, although the theoretical framework of Newtonian mechanics contradicts the Aristotelian framework, there are points of contact with regard to the description of certain patterns of events.

It is not difficult to see the details of this. In cases where a body moves through a medium that offers resistance to its motion, the Newtonian law of motion $F = m \cdot a$ must be supplemented by a friction term so that it becomes $F = m \cdot a + Rv$, where R quantifies the strength of the friction. This equation can be solved for the velocity v, and it turns out that the solution tends toward uniform motion as time progresses.[1] If the friction is substantial, this limit of uniform motion is reached quickly; the final velocity, which remains constant, is $\frac{F}{R}$. This is exactly what the Aristotelian theory predicts. So in situations where significant friction counteracts the accelerating force, the fundamental Newtonian mechanism of force causing acceleration is obscured and it appears that force is responsible for velocity rather than acceleration. Under these special circumstances, Aristotelian relations emerge as an approximation to what is predicted by the laws of Newtonian physics.

The existence of this kind of continuity is to be expected, because Newtonian mechanics must reproduce the empirical successes of Aristotelian mechanics. Is there anything more profound to be discovered in the continuity between Aristotelian and Newtonian mechanics? Can this continuity be used to argue that Aristotelian mechanics contained a kernel of truth that Newton managed to preserve? In a trivial sense, the answer might be yes. Aristotle correctly identified certain phenomenal regularities, and these regularities were preserved by Newton's theory. This shared part could be thought of as a preserved piece of approximate truth. However, this approximate preservation of patterns is at the level of regularities in phenomena and does not represent the kind of truth preservation that scientific realists are usually after. Realism, as commonly understood, is about the discovery of basic causal factors and mechanisms in the physical world, which, accumulative realism claims, we approach ever closer through continuous and incremental improvement of our scientific theories. From this perspective, Aristotle's physics is a disaster. It fails to identify any mechanisms of motion that can be said to be retained, refined, and elaborated in classical mechanics.

3 Emergence and theory change

Emergence can be defined as the appearance of unexpected but robust patterns of behavior within certain application regimes of a theory, usually related to limiting situations of large mass, time, or length scales, or large numbers of degrees of freedom. Emergent patterns differ from the typical behavior determined by the fundamental principles of the underlying theory. Therefore, emergent behaviors, structures, or patterns need additional specific information for their explanation beyond just the principles of the given theory. This additional information may include the number of particles, temperatures, mass and length scales, boundary conditions, and the desired accuracy of the description. Coarse-grained patterns in macroscopic quantities, which differ significantly from the fine-grained, microscopic behavior primarily addressed by the underlying (sub)microscopic theory, provide numerous examples of emergent phenomena.

---

[1] The solution is $v(t) = \frac{F}{R} + b \cdot e^{-\frac{R}{mt}}$, with b a constant and t the time.

The macroscopic gas laws are a case in point. At the macroscopic level, characterized by large numbers of particles and temperatures typical of our everyday environment, the behavior of gases is relatively simple and can be characterized by regularities in a small number of quantities (pressure, temperature, and volume). But sub-microscopically, gases are systems with many particles that generally do not behave in a simple orderly way at all.

More generally, the basic ontology of a theory, together with its fundamental laws, produces descriptions with a broad scope of application. However, emergence leads to effective descriptions that possess only approximate validity within specific and limited domains of application of the theory. The patterns that characterize these effective descriptions function as the "laws" of *effective theories*. From the perspective of basic theory, these are merely contingent regularities between non-fundamental and sometimes even nonexistent quantities.

Evidently, when a successful scientific theory is replaced by a new one, the new theory must be able to reproduce the successes of the first theory. For example, the successes of phenomenological thermodynamics are reproduced by statistical mechanics, and the successes of classical mechanics are reproduced by the theory of relativity and by quantum mechanics. Even the successes of Aristotelian mechanics are reproduced by classical mechanics, as we have seen. What all these cases have in common is that the old successful predictions are not exactly reproduced, but only approximated; strictly speaking, the old predictions are falsified. Moreover, from the point of view of the new theories, the old successful patterns are only conditionally valid, depending on conditions that define a narrow sub-domain of the theory's application. The old successes appear as emergent patterns, part of effective and non-fundamental descriptions.

The occurrence of emergence in the transition from one theory to the next suggests that the relationships between successive theories are usually not about refinement or incremental improvement, but involve the discovery of new conceptual frameworks not previously anticipated. Therefore, emergence challenges the accumulative realist assumption of a gradual increase of truth or approximate truth.

Even concepts that are absolutely fundamental and central in a physical theory can prove to be of mere effective and pragmatic value when the theory is replaced by a new one. A recent example of this is provided by the disappearance of the notion of an *object*, a *thing* possessing individual identity, in the transition from classical physics to quantum theory.

4 Classical particles as emergent entities

The world of classical physics, like the world of our direct experience, is a world of *objects*, *things*. Objects have definite physical properties, like position and velocity, and have definite histories by means of which they can be followed over time. In classical mechanics, the typical object is a *particle*—a notion that is central to the theory. No two particles can ever occupy the same position, so particles can always be told apart on the basis of where they are; moreover, each particle can be reidentified over time by means of the path it follows. Thus, classical particles, like the objects of everyday experience, are individuals.

Surprisingly, this notion of an individual object with definite properties is hard to reconcile with quantum physics.[2] According to relativistic quantum field theory, it is impossible to have a physical system that with certainty will be found within a spatial domain of a given finite extension (see, e.g., Halvorson and Clifton 2002, Dieks2023b). Therefore, the physical "things" that are allowed by relativistic quantum field theory cannot be localized objects. A further unexpected result is that even if we try to think of particles as non-localizable and non-classical entities, the so-called Unruh effect shows that their presence will generally be observer-dependent. For example, if an inertial observer measures a vacuum, without particles, an accelerated observer may find evidence showing that there are particles after all (Wald 1994, Ch. 5; Halvorson and Clifton 2002). This is obviously difficult to reconcile with the picture of particles as entities whose existence is objective and independent of observation.

Despite these and other seemingly bizarre results, it is clear that quantum physics should be able to make contact with the world of daily experience. The classical particle concept must become effectively applicable when transitioning from the quantum to the classical world (Dieks and Lubberdink 2020, Dieks 2023a). Indeed, there is a limiting regime of quantum theory, characterized by large masses and many environmental degrees of freedom, where typical quantum effects become difficult to detect. In this specific and limited domain, quantum mechanisms are hidden from view and the world may appear classical.

In particular, patterns in events will arise that create the impression of particle-presence. Although this happens in a very tiny corner of the total application domain of quantum mechanics, it is a corner with great significance for humans in their daily lives. But even within this classical regime, the particle picture will only work if no sophisticated experiments are performed that are able to reveal quantum effects. Quantum features remain

---

[2]In what follows we use standard interpretative ideas concerning quantum theory. There exist alternative interpretations with different roles for the notion of a particle. This situation complicates the predicament of the realist: the different interpretations are empirically equivalent, but they cannot all be true. Do some of them achieve their empirical success by some miracle? This underdetermination of theoretical structure by empirical data forms an important part of the argument against the cogency of realism, but we cannot go into this part of the argument here.

present in principle, and their detection can prove the classical particle picture incorrect.[3]

The situation resembles that of Aristotelian versus modern physics. As long as we do not make accurate measurements and stay within our usual everyday conditions, there seems nothing wrong with Aristotelian mechanics. But if we get precise and also look at what happens in unusual scenarios, we must conclude that reality is very different from what it seems.

5 Emergence and continuity

In the transitions from Aristotle to Newton and from classical to quantum there is certainly continuity. In both cases, old regularities are derivable from the new theory as effective descriptions, approximately valid in a small part of the new theory's domain. This may seem to confirm the continuity expectations of adherents of accumulative scientific realism, who claim that continuity is a consequence of truth preservation.

However, the example of Aristotelian mechanics as a limiting case of Newtonian theory should give us pause. There is only a small class of phenomena for which Aristotle's theory yields predictions close to the Newtonian ones. Within this domain, the emergent pattern derivable from Newton's theory is on the level of events but does not extend to mechanisms, causal links, and explanations. Aristotle's framework revolving around such concepts as natural places, natural versus forced motion, $v = \frac{F}{R}$, stands in such strong contrast to the Newtonian account that Aristotle's mechanics is often not even considered to be a part of science at all. None of the principles of motion used by Aristotle was taken over by Newton. From this perspective, the transition from Aristotle's theory of motion to classical mechanics does certainly not support the claim of truth preservation.

Nonetheless, there are phenomena within the scope of Newton's theory that can also be accommodated by Aristotelian mechanics. Doesn't this overlap cry out for explanation, and isn't the only reasonable explanation a common element of underlying truth, as suggested by the no-miracle argument? The answer is 'no'. There is an obvious alternative explanation for the continuity between Aristotle and Newton, one that does not require a shared kernel of deeper truth. This explanation is simply that Newton's theory has to reproduce the (limited) empirical success of Aristotle's theory—if it were unable to do so, this would constitute a fatal objection to Newton's theory. Realists and anti-realists alike agree that successor theories must be able to reproduce the empirical success of their predecessors. This self-evident demand for the preservation of empirical success is enough to understand that successive theories must have a common part, namely the

---

[3] In fact, important progress has been made, during the last decades, in showing that seemingly macroscopic objects are actually quantum.

set of observable regularities covered by both theories. Aristotle and Newton were both able to describe bodies moving through a medium that offers resistance.

This existence of continuity on the level of observable phenomena is to be expected *a priori*, independent of realism or empiricism. What is more, even empiricists will expect a continuity that goes deeper than just the preservation of success at the level of the directly observable. This is because scientific theories do not contain, within their conceptual frameworks, any built-in demarcation line between descriptions that apply to what is observable by humans and descriptions of things unobservable to humans. Scientific theories have the form of objective descriptions that do not refer to observers or human perception. Therefore, it is to be expected that assertions valid for observable things and processes will also extend, at least to some extent, to proccesses and events that defy direct human observation (for example, because they are about objects that are too small to be seen). Thus, Aristotle's theory of motion predicted not only that observable heavy objects fall (striving as they are to reach their natural places) but also that invisibly small grains of heavy material will do the same. This absence of a dividing line between the observable and the unobservable applies to the conceptual frameworks of all scientific theories. Therefore, if a successor theory is able, as it must be, to reproduce the observable regularities successfully predicted by a predecessor, it should be expected to reproduce also the predictions of the old theory in a regime going beyond what is directly observable. In the example of Aristotle and Newton, the set of nearly identical predictions thus includes not only certain motions of observable bodies but also motions of unobservable objects.

Therefore, the fact that new theories are able to explain the successes of their predecessors, as emergent patterns both concerning the observable and parts of the unobservable, does not automatically imply that a piece of truth concerning the workings of nature has been preserved.

6 Emergence and truth

Accumulative realism claims that our empirically successful theories must possess a good deal of partial and approximate truth; how else could their success be explained? A considerable part of this truth comes from earlier successful theories, and these truths will be carried over again to future theories. Accordingly, we can be pretty sure that principles, processes, and entities that have withstood all theory change to date represent pieces of truth that will remain unaffected by future theoretical developments (cf. Vickers 2022). But as we have argued, there are reasons to doubt this view or at least to put it into perspective: typically, new theories transform older schemes into effective descriptions that are only approximately valid within limited portions of the new theories' domains. Laws, principles, and mechanisms of new theories may well be radically different from the old ones. In such cases, there is no preserved truth at the level of laws, causality, and explanation. Even a basic concept like 'particle', which survived theoretical change for so many centuries, has turned out to be ephemeral.

In conclusion, accumulative scientific realism in the form we have discussed does not seem a viable account of scientific progress. The history of science shows that the empirical success of a theory may well be explainable from principles and mechanisms that are radically different from the explanatory devices offered by the theory itself.

References

Alai M (2021) The Historical Challenge to Realism and Essential Deployment. In: Lyons TD and Vickers P (eds) Contemporary Scientific Realism: The Challenge from the History of Science, Oxford University Press, Oxford, pp 183–215, doi: 10.1093/oso/9780190946814.003.0009

Cordero A (2011) Scientific Realism and the Divide et Impera Strategy: The Ether Saga Revisited. Philosophy of Science 78: 1120–1130, doi: 10.1086/662566

Chang H (2003) Preservative Realism and Its Discontents: Revisiting Caloric. Philosophy of Science 70: 902–912, doi: 10.1086/377376

Dieks D (2023a) Quantum Individuality. In: Becker Arenhart JR, Wohnrath Arroyo R (eds) Non-Reflexive Logics, Non-Individuals, and the Philosophy of Quantum Mechanics. Synthese Library 476, Springer, Cham: chapter 2, doi: 10.1007/978-3-031-31840-5_2

Dieks D (2023b) Emergence and Identity of Quantum Particles. Philosophical Transactions A, doi: 10.1098/rsta.2022.0107

Dieks D (2023c) Emergence, Continuity, and Scientific Realism. Global Philosophy 33:44, doi: 10.1007/s10516-023-09696-w

Dieks D, Lubberdink A (2020) Identical Quantum Particles as Distinguishable Objects. Journal for General Philosophy of Science, doi: 10.1007/s10838-020-09510-w

Halvorson H, Clifton R (2002) No Place for Particles in Relativistic Quantum Theories? Philosophy of Science 69: 1–28, doi: 10.1086/338939

Laudan L (1981) A Confutation of Convergent Realism, Philosophy of Science 48: 19–49, doi: 10.1086/288975

Psillos S (1994) A philosophical study of the transition from the caloric theory of heat to thermodynamics: Resisting the pessimistic meta-induction. Studies in the History and Philosophy of Science 25: 159–190, doi: 10.1016/0039-3681(94)90026-4

Psillos S (2009) Knowing the Structure of Nature: Essays on Realism and Explanation. Palgrave/MacMillan, London, doi: 10.1057/9780230234666

Psillos S (2022) Realism and Theory Change in Science. In: Edward N. Zalta and Uri Nodelman (eds.), The Stanford Encyclopedia of Philosophy (Fall 2022 Edition).

Vickers P (2022) Identifying Future-Proof Science. Oxford University Press, Oxford.

Wald R (1994) Quantum Field Theory in Curved Spacetime and Black Hole Thermodynamics. University of Chicago Press, Chicago

Worrall J (1989) Structural Realism: The Best of Both Worlds? Dialectica 43: 99–124, doi: 10.1111/j.1746-8361.1989.tb00933.x

Justifying scientific beliefs: an anti-naturalist and anti-pragmatist perspective

Michel Ghins

Institut Supérieur de Philosophie, Université Catholique de Louvain, Place Cardinal Mercier 14/L3.06.01, 1348 Louvain-la-Neuve, Belgium
E-mail: `michel.ghins@uclouvain.be`

> **Abstract.** This paper defends epistemological scientific realism (ESR), understood as the philosophical position that we have stronger reasons to believe in certain claims about some unobservable entities (conceived as sets of properties) posited by our best scientific theories, rather than withholding judgment about their existence. Following a critique of explanationist defenses based on the "no miracle argument", I propose an empiricist inductivist justification of ESR, drawing parallels with how we justify belief in claims about immediately observable entities in everyday experience. This justification is non-naturalistic, as it is not grounded in the history or practice of science, and it is normative, providing a framework to evaluate the strengths and weaknesses of arguments supporting belief in the reality of entities posited by our best scientific theories.

Today, most empiricists embrace some form of naturalism, viewing philosophy as a scientific discipline. They argue that philosophical claims about science should be grounded in the study of scientific practice. Henk De Regt's book *Scientific Understanding* exemplifies this approach, as he states:

> The aim of the book is to develop and defend a theory of understanding that describes criteria for understanding actually employed in scientific practice. (De Regt 2017, 6)

However, I contend that the role of philosophy is not to describe how scientists, past or present, conduct their work. Philosophy is not a form of metascience whose empirical basis is the practice of scientists. As Bas van Fraassen rightly observes, the aim of philosophy is not to describe facts.

> Do electrons exist? Are atoms real? These are not philosophical questions. Whether electrons exist is no more a philosophical question than whether Norwegians exist, or witches, or immaterial intelligences. Questions of existence are questions about matters of brute fact, if any are, and philosophy is no arbiter of fact. (van Fraassen 2017, 95)

It is the responsibility of scientists, not philosophers, to argue for the existence of entities such as electrons, mitochondria, and tectonic plates. In the debate on scientific realism, philosophers are concerned with the normative task of evaluating the strength of the kinds of arguments that are claimed to justify belief in the reality of certain entities posited by scientific theories and the truth of statements about them. This examination of the solidity—or lack thereof—of various types of arguments is an epistemological and logical endeavor. It deserves careful scrutiny because it directly impacts the level of confidence in our best scientific theories.

While the arguments supporting belief in the existence of electrons obviously differ in contents from those justifying belief in mitochondria, some arguments are stronger than others. Philosophers are not primarily interested in specific arguments but in the grounds for the strength of various types of arguments. According to epistemological scientific realism, we have stronger reasons to believe in the truth of scientific claims about some entities inaccessible to direct observation, rather than suspending belief about them. Thus, philosophers face the challenge of identifying the grounds for the cogency of the arguments supporting belief in the existence of entities inaccessible to direct observation.

Most scientific realists rely on a form of reasoning known as "inference to the best explanation" or "abduction" to justify belief in the existence of entities that cannot be directly observed. In the following discussion, I will argue that explanationist strategies fall short of providing sufficient reasons to believe in the existence of certain indirectly observed entities. Instead, I will propose an alternative inductivist strategy that aligns more closely with a moderate version of empiricism and offers robust support for a selective version of epistemological scientific realism.

1 Inference to be the best explanation: A critique

Explanationist strategies that rely on *Inference to the Best Explanation* (IBE), or abduction, as a method for justifying true beliefs have a long-standing role in philosophical reasoning. A typical formulation of IBE can be presented as follows:

1. F is a fact.

2. Hypothesis H explains F.

3. No other available hypothesis explains F as well as H.

Conclusion: H is true.

One of the most famous examples of IBE is Putnam's "No Miracle Argument" (NMA). In this argument, Putnam (1978, 18) asserts that the success of scientific theories in making accurate empirical predictions would be inexplicable—that is to say, *miraculous*—unless we assume that their claims about entities like electrons, genes, and other theoretical posits are at least partially true. His argument can be summarized as follows:

1. Fact F^*: Theory T makes accurate predictions.

2. Hypothesis H^* (Theory T is partially true) explains F^*.

3. No other available competing hypothesis explains F^* as well as H^*.

Conclusion: H^* is true.

It is well-known that such abductive arguments are *logically invalid* since it is always possible that some unknown alternative hypothesis—one incompatible with H—could explain the fact F better. (Stanford 2006) This challenge raises a critical question: Can the No Miracle Argument be made valid, and even sound, by introducing an additional premise? This would be analogous to the approach of making inductive reasoning logically valid by assuming the uniformity of nature.

Alan Musgrave proposed to add the following premise 1' to the No Miracle Argument to get:

1'. It is reasonable to believe that the best explanation of any fact is true.

1. F^* is a fact.

2. Hypothesis H^* explains F^*.

3. No available competing hypothesis explains F^* as well as H^*.

Conclusion: It is reasonable to believe that H^* is true. (Musgrave 2017, 80)

Notice the epistemic shift in Musgrave's version of the No Miracle Argument (NMA), marked by the inclusion of the phrase "it is reasonable to". This addition significantly weakens the position of scientific realism. A scientific realist cannot rest satisfied with the claim that believing in entities like electrons is *not* irrational, since anti-realists like van Fraassen already concede this point. Instead, the realist must go further, showing that it is *more rational* to believe in the existence of entities such as electrons, etc. than to deny their existence or remain agnostic about them.

While Musgrave's reformulation undoubtedly makes the argument deductively valid, does it make the argument sound? Specifically, do we have sufficient justification to believe in premise 1', which asserts a strong connection between the best explanation and its truth, or at least that this connection is more likely to hold?

Following Lipton (2004), we must distinguish between two types of explanations, similar to the distinction between valid and sound arguments. A sound argument is a valid argument in which all premises are true. Analogously, a *true* explanation is a satisfactory explanation in which all premises are true. An explanation is considered satisfactory, or "lovely," if it provides a good understanding, though it may still be false. For instance, Ptolemy's theory of crystalline spheres provides a clear and understandable explanation of the stability of planetary motions. We understand that if planets are attached to hard regularly rotating spheres, their trajectories remain unchanged over time. This is what we observe, at least for short periods of time. However, Ptolemy's hypothesis is false, and so is his explanation.

Proponents of *Inference to the Best Explanation* (IBE) make the distinction between lovely and true explanations since multiple satisfactory explanations of the same phenomena can be provided. By implementing a top-down strategy, the friends of IBE assess the *internal* merits of competing explanations to conclude that the loveliest explanation—the one with the greatest explanatory power—is true or at least more likely to be true.

However, even if we could all agree on the criteria for comparing the explanatory power of competing hypotheses, and even if all possible explanations for a set of data were available (a highly unrealistic scenario), there would still be no stronger reasons to believe that the most satisfactory explanation is true. Why? Because *the explanatory power of a hypothesis does not, in itself, justify belief in its truth.* (Ghins 2024, 61)

This is the fundamental difficulty with any explanationist strategy for justifying beliefs. It may not be irrational to believe in the truth of hypotheses that correctly predict and explain phenomena. But is it *more rational* to believe that nature is organized according to what we deem to be a good explanation, according to our standards of understanding and intelligibility, even if those standards were universally shared or rooted in human nature? What connection to reality could possibly be guaranteed by the explanatory power of a hypothesis, given that this power is evaluated on the basis of subjective criteria for understanding? In my view, none.

As Peter Lipton asks: Is the loveliest explanation—the one that pleases us most (and thus is the most satisfactory to us)—also the likeliest to be true? (Lipton 2004, 61) Is there a pre-established harmony between our explanatory preferences and reality? Such a Leibnizian harmony between reality and the explanatory requirements of our minds *might* exist, but how could we possibly argue for it convincingly? This is what Lipton refers to as "Voltaire's objection."

> (...) supposing that loveliness [of an explanation] is as objective as inference (...) What reason is there to believe that the explanation that would be the loveliest, *if it were true* [emphasis is mine], is also the explanation that is most likely to be true? Why should we believe that we inhabit the loveliest of all possible worlds? (Lipton 2004, 70)

Unlike Lipton, I do not believe that an explanation would be the loveliest "if it were true" or correct. An explanation can be the loveliest or most satisfactory even if its premises are false, and vice versa. I wish to keep the concept of *loveliness* entirely separate from *truth*. However, I do agree with Lipton on the main point: there is no intrinsic connection, let alone harmony, between the beauty or elegance of an explanation and the actual facts of the world. To believe otherwise is to fall into an idealist prejudice, which assumes that our requirements for understanding grant us privileged cognitive access to an external reality.

Bas van Fraassen rightly emphasizes that there is no relationship between explanatory power and truth. If this is correct, then the No Miracle Argument cannot be salvaged, even if we accept that scientific realism is the only acceptable—and therefore the best—explanation for the predictive success of scientific theories. The epistemic gap between explanation and truth makes such a rescue impossible. Abductive arguments may play a valuable heuristic role in generating new explanatory hypotheses, as Peirce highlighted, but their explanatory appeal provides no justification for believing them to be true.

2 Induction as an alternative to abduction: An example

To defend epistemological scientific realism, I propose an inductivist strategy as an alternative to explanationist arguments. In cases where the existence of entities is inferred, the strength of abductive arguments stems from hidden deductions involving causal propositions that are empirically and inductively justified. I will illustrate this with a simple example of *Inference to the Best Explanation* (IBE) discussed by van Fraassen (1980, 19–20):

1. It is more reasonable to believe that the loveliest explanation of any fact is true.

2. Fact F: Grey hair lies on the floor, cheese disappears, and specific little noises are heard.

3. The presence of a mouse (H) explains F.

4. No available competing hypothesis explains F as elegantly as H.

Conclusion: It is more reasonable to believe that H is true, i.e., that a mouse is present.

The persuasiveness of this argument does not come from its abductive character. Rather, if it convinces us, it is because it relies on premises—grounded in induction—that are not explicitly stated, as I will now attempt to show. To avoid circularity, we must first define what a mouse is: by definition, a mouse is an animal with four legs, a long tail, small ears, and a pink snout.

In addition—and this is the crucial point—constant causal correlations between certain events have been observed. In general, causal connections can be established through empirical methods, such as those codified by John Stuart Mill (1843), provided they are sufficiently refined. Causal relations are asymmetrical: causes produce effects, not the other way around. However, once a causal relation has been empirically ascertained, we obtain an "if and only if" logical connection between cause and effect: if the cause occurs, the effect follows, and vice versa. These previously identified causal relations allow us to infer the presence of a cause based on the observation of its effects.

In the current example, finding grey hair on the floor and the disappearance of cheese regularly coincide with the shedding of grey hair and the consumption of cheese by an animal. Through repeated observations, we have learned that mice (as defined earlier) are the only creatures that exhibit these properties (such as grey hair loss) which are causally correlated with the observed effects (grey hair on the floor, missing cheese, etc.). By inductive reasoning, we conclude that the occurrence of these observed effects generally implies the presence of a mouse, which is the cause.

In this particular case, based on the presence of grey hair and missing cheese, we can infer—perhaps to our dismay—that there is at least one mouse in the house. Thus, we have indirectly detected the presence of a mouse by inferring its existence from the evidence, even though we haven't observed it directly. Later, with patience (or luck), we might observe a mouse directly, which would strengthen our belief in its presence, as we would immediately see a larger set of its properties.

The abductive reasoning in this example is supported by the following deductive argument:

1. Facts (F): Grey hair on the floor, disappearance of cheese, specific little noises.

2. Inductively confirmed hypothesis (H): These facts (F) are caused by the presence of animals with grey hair that eat cheese, etc.

3. Inductively confirmed association (A): Mice, defined as four-legged animals with specific characteristics, also shed grey hair and exhibit other associated properties.

4. Alternative hypotheses (e.g., a rat, a mischievous neighbour) are not confirmed by the observations.

Conclusion: It is more reasonable to believe that a mouse is present.

This argument is deductively valid, but it rests on premises whose truth has been established through induction, particularly premises 2 and 3. Why, in addition, do we end up with a "lovely" explanation of the observed facts? Because hypothesis H describes causal processes. The shedding of hair, making noise, and eating cheese are empirically verified causal events—sequences of properties that unfold over time.

Even if we grant that the presence of a mouse best explains the available evidence, this does not necessarily mean it is more rational to believe in the mouse's presence than to suspend our judgment. If we are justified in concluding that a mouse is present, it is because of previously verified causal processes, which enable us to trace back the existence of the cause (the mouse) from its effects (the grey hair, the missing cheese). These causal processes, in turn, form the basis for explaining the empirical evidence.

In summary, the argument for the presence of the mouse is a logically valid deductive argument. If its premises are true, then the argument is sound. By adding the premise that a mouse is present and taking a description of the observed facts as a conclusion, we arrive at a correct explanation of the factual evidence.

One might object that while the argument presented is valid, it is not sound, as we must establish in this particular instance that the alleged causal connection holds—specifically, that premise H is true. Indeed, it is possible that another cause could explain the observed facts. For example, a malevolent neighbour could be playing tricks by deliberately placing grey hair on the floor and creating other clues. Such alternative explanations are often hypothesized and then discarded in abductive reasoning as less elegant or "lovely." However, I argue that we should rely solely on empirical evidence rather than the subjective appeal of an explanation. If there is no empirical evidence to support the hypothesis of a mischievous neighbour, we have no reason to entertain that possibility. Observational evidence of external facts is far more reliable than the supposed internal virtues or elegance of competing explanations.

But what about the possibility of alternative causes that have not yet been conceived? (Stanford 2006) Inductivists need not be overly concerned with these, as unknown alternatives cannot be empirically tested or inductively confirmed. The mere possibility of unconceived alternative explanations does not undermine the evidence we currently have, which provides stronger reasons to believe in the presence of a mouse. However, we must acknowledge that since premise H is not established with certainty, we should treat it as only having a higher probability of being true.

3 Direct and indirect observation

In the previous section, I have shown that belief in the instantiation of properties is justified when we rely on deductive arguments whose premises describe causal connections, and which have been inductively confirmed. Notice that such confirmation is possible because, in the previous example, we were dealing with immediately observable properties. But can this approach be extended to properties that are only accessible through instruments, such as telescopes, microscopes, and other observation devices?

In addition to directly observable properties—like hardness, roundness, or hairiness—I also include in the category of observable properties certain scientific properties, such as mass, charge, and temperature. However, I exclude properties like internal spin, strangeness, and charm from this category. Some philosophers may rightly object that terms like "mass," "charge," and "temperature" belong to a theory-laden language. Moreover, the meanings of these terms have only become clear and stabilized through a long and painstaking historical process. However, once we have grasped the meaning of a term like "gravitational mass," we can readily verify that my teacup is heavier than my pen through direct observation. Similarly, once the meanings of "positive charge" and "negative charge" are understood, we can empirically verify the presence of charges of the same sign (positive or negative) by directly observing the repulsion of thin leaves in an electroscope. Although the presence of charges may initially have been hypothesized through abductive reasoning—the heuristic value of which I do not dispute—only observation can support belief in their instantiation.

Critics might immediately object that there is a distinction between observing a property P and observing *that* something possesses the property P. For instance, observing the property of hardness is not the same as observing *that* an object is hard. Actually, this distinction has little bearing on the issue of realism, since the truth of propositions and the instantiation of properties are closely connected. Is it true that there is a hard object on my desk? The truth of this statement depends on a fact: the instantiation of the property of hardness, which is confirmed through direct perception. Similarly, is it true that the gravitational mass of my teacup is greater than that of my pen? This assertion, too, can be verified or falsified through direct observation.

According to my empiricist stance, no property is cognitively accessible unless it is observable by us, either directly or indirectly. However, I include in the category of observable properties some scientific properties, such as charge and gravitational mass, which are not considered observable by most empiricists. These properties, like many other properties in science, can assume various continuous or discrete values and are referred to as *determinable properties* because they can take on specific determinate values.

Due to the limitations of our senses, we cannot directly perceive very large or small values of mass, charge, volume, velocity, and similar properties. However, since we can observe some values of these properties directly, I submit that very large or small values of them can still be considered observable in a broader sense. Even though an extremely high velocity isn't directly perceivable, it is still a velocity and thus resembles directly observable velocities. While this expanded notion of observability is not consistent with strict empiricism, such extension is justified because resemblance allows us cognitive access to similar properties through detections whose reliability is supported by empirical induction, as I will show below.

Now, let us turn to properties that are unobservable in principle, which I refer to as *purely theoretical properties*. These properties are beyond the reach of any possible observation—either direct or indirect—not only in practice but in principle. In this sense, they are transcendent. Purely theoretical properties, which do not resemble anything accessible to perception, are common in elementary particle physics. Examples include internal spin, strangeness, and charm. Unlike properties such as volume or mass, these cannot be verified through ordinary sensory experience. Therefore, we are never justified in believing in the instantiation of such purely theoretical properties.

4 Four conditions for justified belief in the instantiation of properties

The primary challenge faced by epistemological scientific realists is justifying belief in properties that, while directly unobservable due to practical constraints or perceptual limitations, are still detectable. The first condition for believing in the reality of such properties is that they must not only be observable in principle but they must also have been actually observed. This leads us to formulate the following observation condition:

Observation Condition (O): *To have stronger reasons to believe in the existence of a property rather than to suspend judgment or disbelieve, it is necessary for that property to be either directly observed or indirectly observed through detection.*

In scientific observation, sight holds a privileged status, and various detection instruments enhance its capabilities. For example, consider ordinary eyeglasses, used by those with impaired vision. Who would argue that a farsighted person's observations of distant objects are less credible simply because they use glasses rather than relying on unaided vision? Now, let us consider more powerful optical devices, such as telescopes. Inside a telescope, we directly see what we typically refer to as "images" with specific properties, such as geometric shapes. Geometric shape is a directly observable property of celestial objects, such as planets. If we were close enough to a planet, we could directly observe its approximately spherical shape.

Thus, we commonly say that we "observe" a planet through a telescope. In fact, what we directly observe are the properties of the image inside the telescope, but the shape of the image (A) corresponds to the shape of the planet (B) through a logical *iff* (if and only if) relation: if A, then B, and conversely. Moreover, B causes A. Thus, according to my terminology, the telescope allows us to indirectly observe or *detect* the shape of the planet.

These remarks can be extended to various types of microscopes and telescopes, which permit us to see entities such as viruses and distant galaxies. When direct observations and those made with the aid of a microscope agree, we can consider the microscope reliable—at least within the overlapping domain of these observations. By induction, we then extend the microscope's reliability to properties that are not immediately visible. Furthermore, our knowledge of the laws of optics, verified through induction in the realm of directly observable properties, justifies trusting the microscope when detecting properties invisible to the naked eye. Gradually, through inductive reasoning, we expand the domain of accessible observable properties to increasingly broader realms of detection. For this reason, it is legitimate to regard very large or very small values of these properties as "observable" in a broad sense, even if they are only detectable.

It is important to note that it is not always necessary to know the causal laws underlying the workings of an instrument in order to trust its results. For example, the ancient Romans used polished lenses to correct vision, despite being unaware of the laws of refraction, let alone electromagnetism. Similarly, Galileo and his contemporaries knew very little about the inner workings of the *canocchiale* (telescope). Nonetheless, when close-range observations of an object, such as a ship, matched those made from a distance using the *canocchiale*, they could empirically confirm a causal connection between the properties directly observed through the telescope and the detected properties of the distant object. Even the Aristotelians, who at first were skeptical, quickly acknowledged the reliability of Galileo's telescope.

This inductive approach, which Philip Kitcher calls the "Galilean strategy" (2001, 173–174), can also be applied to other instruments, such as the microscope. (However, unlike Kitcher, I do not believe the Galilean strategy can be applied to purely theoretical properties.)

In cases of indirect epistemic access to properties, we rely on the causal relationships between the properties being detected and those that are directly observed. The verification of these causal relationships depends upon previously confirmed inductive generalizations. By knowing these relationships, we can trace back the causes (the detected properties) from their effects (the directly observed properties).

In everyday experience, what justifies us in asserting the existence of objects such as a teacup or a rose is their immediate presence in perception, provided good observational conditions obtain. This perception is supported by underlying causal connections, which, although we may not fully understand them, we have stronger reasons to believe that such causal links exist, based on Mill's rules. For instance, when we manipulate a perceived object in certain ways, we observe systematic changes in how its properties are perceived, further reinforcing our belief in the presence of a causal link.

As we saw earlier, in order to check the reliability of a new instrument, we compare its measurements with those obtained with an already established, accurate instrument in the same empirical domain. If the results from both instruments concord in such overlapping domain, we inductively extend the reliability of the new instrument to broader detection domains. This method follows what Kitcher calls the "Galilean strategy" I mentioned above. Step by step, through this methodical process we justify the significant expansion of the range of detectable properties made possible by the invention of new instruments and measuring devices.

What is more, when we have detailed knowledge of the empirical causal laws governing the mechanisms that underlie the functioning of instruments or observation devices, we have strong grounds to believe that the causes of the observed effects possess certain specific properties. These causes—such as mass, charge, or velocity—are instantiated properties that, while not directly observed, can be judged to have been detected. From this, we can establish a second condition that must be met to hold a justified belief in the existence of detected properties.

Causality Condition (C): *To have stronger reasons to believe in the existence of a property that is not directly observed, this property must be detected—i.e., empirically verified as causally linked to properties that are directly observed through the use of reliable instruments.*

To reinforce my belief in the instantiation of a property, I can mobilize several perceptual modalities and check whether they give concordant results. For example, to confirm that an object on my desk is hard, I can touch it, strike it to hear the characteristic sound of a hard object, and observe its visual properties that suggest hardness. Here, three distinct perceptual modalities—touch, hearing, and sight—come into play, each functioning independently. Each modality provides empirical access to the property of hardness.

Within each modality, I can repeat observations in various ways to ensure that the results are consistent. For instance, through different forms of touch, I consistently confirm the presence of hardness. Likewise, though the sounds I hear vary slightly with each strike, they all consistently indicate hardness.

Additionally, I can apply similar methods to verify other properties that distinguish the object as a teacup rather than a vase or another type of item.

These observations suggest the need for an additional requirement—an invariance condition—to justify belief in the existence of directly observed properties:

Invariance Condition (Ia): *To have stronger reasons to believe in the existence of a directly observed property, it is necessary and sufficient that repeated observations of the property, through distinct and independent perceptual modalities, yield invariant results, at least approximately.*

For a directly observed property, the invariance condition is both necessary and sufficient to justify belief in its instantiation. There is no doubt that this condition of invariance is rooted in the truth of generalizations describing Millian causal connections between the perception of property (under favorable conditions) and the actual instantiation of that property. Perception is known to be a complex process involving causal links—still not fully understood—between external properties and the properties of our sensory organs, nervous system, and brain.

When different observations, relying on distinct causal pathways, yield consistent results, our confidence in the reality of a given property increases. Why? Because previous experience has shown that this approach minimizes the risk of error. Over time, we have learned that beliefs supported by such a procedure are less likely to be falsified. Indeed, when we seek to resolve doubts about the properties attributed to an entity, we repeat and vary our observations. This method, again, is justified by induction.

It is important to emphasize that our belief in the instantiation of a property (or set of properties) is not based on an argument that the property best explains the concordance between different perceptions. This is *not* an inference to the best explanation of the agreement of various observations. Rather, in each perception, the property is directly observed. Repetition simply provides new instances of perceiving the same property, and the consistency of these observations reinforces the stability of our beliefs. This stability arises because, through induction, we have learned that beliefs strengthened in this manner are more resistant to potential falsification.

What can we now say about indirectly observed, or detected, properties? To justify belief in the existence of such properties, we must empirically verify that causal connections exist between directly perceived properties (clues) and the detected properties. However, by analogy with the invariance condition for directly observed properties, we must also require that repeated detections using different empirical methods yield consistent results.

Invariance Condition (Ib): *To have stronger reasons to believe in the existence of a detected property, it is necessary that repeated detections of this property, using distinct and independent empirical methods, yield invariant results, at least approximately.*

For detected properties, this condition is necessary but not sufficient, as the distinct empirical methods must also be reliable. This reliability is grounded in the causal requirement outlined earlier. Some methods of detection, such as those used in astronomy, involve instruments, while others, like the detection of a mouse, may not.

In many scientific contexts, determining the exact value of a detectable property is impossible without the aid of measuring instruments. Therefore, we must introduce an additional condition: the measurement condition.

Measurement Condition (M): *In the quantitative sciences, to have stronger reasons to believe in the existence of a detected property with a specific value, it is necessary for the property to be quantitatively measured using instruments whose reliability has been previously and independently established.*

Together, we now have four conditions for justified belief: Observation, Causality, Invariance, and Measurement. For brevity, I will refer to these as the OCIM conditions. The satisfaction of all four OCIM conditions is both necessary and sufficient to justify belief in the instantiation of a detectable property. These conditions—crucially the causality condition—allow us to ascend, in a bottom-up approach, from directly observed properties to the properties that cause them.

On the other hand, there are never strong reasons to believe in the instantiation of purely theoretical properties. Why? Simply because such properties transcend any empirical cognitive access. For empiricists, they are beyond cognitive access *tout court*. These properties are epistemically transcendent. It is impossible to empirically verify that purely theoretical properties are causally connected to observed properties. Regarding such theoretical properties, I recommend adopting an agnostic stance: while these properties might exist, we will never have compelling evidence to believe in their reality.

We can now summarize these four conditions as follows:

Requirement R: *To have stronger reasons to believe in the existence of a detectable property, it is necessary and sufficient that this property has been detected multiple times using various methods, whose reliability is grounded in empirically and inductively confirmed causal connections between the detected property and directly observed properties. Furthermore, the results of these observations or measurements must be consistent.*

This requirement is *normative*, meaning that the fulfillment (or lack thereof) of the four OCIM conditions provides a general framework for evaluating the strength or weakness of an argument supporting the existence of a particular entity (which is understood as a set of properties). The scientific realism I propose is thus a decisively normative philosophical stance. Its plausibility does not depend on whether scientists actually use arguments that conform with this requirement when arguing for the existence of specific entities. Instead, the strength of this realist position lies in the similarity between the arguments used to support belief in detectable entities and those used to support belief in ordinary, immediately observable entities.

This form of realism harmonizes with the idea that science is an extension of common sense, a view supported by philosophers such as W. V. O. Quine, who remarked:

> Science is not a substitute for common sense but an extension of it.
> (Quine 1976, 229)

The scientific properties we are justified in believing to be instantiated are either identical or similar to the properties of everyday objects, which we access through direct perception. These properties are all observable in a broad sense. Moreover, the existence of properties posited by a theory can be ascertained as long as their detection is confirmed using procedures akin to those employed in everyday life—that is, through repeated and varied observations.

The adoption of this inductive empirical strategy for defending selective scientific realism offers a valuable alternative to traditional vindications of scientific realism that rely on the no-miracle argument and explanationist strategies. While I do not deny that the prediction of novel and unexpected facts is relevant for reasonable belief in certain parts of a theory, it is important to clarify the nature of that relevance. If the observation of a novel fact provides grounds for believing in specific components of a theory, it is not because such observations evoke psychological feelings of surprise or awe, since these lack epistemological significance, nor because the theory has the capacity to explain the novel fact. Rather, the epistemological power of novel observations stems from the presence of convincing empirical evidence establishing a causal link between specific parts of the theory and the novel fact.

This kind of novelty can also be linked to the invariance condition, when new detection methods enable the connection between theoretical properties and new observations, thereby reinforcing belief in their existence. Classic examples from the history of science illustrate this point, such as the observation of a bright spot in the center of the circular shadow cast by a circular screen (providing additional evidence for the wave nature of light), and the deflection of starlight near the Sun (which supported Einstein's construal of the gravitational field in his general theory of relativity).

5 The case for the existence of Neptune

In this section, I illustrate the bottom-up inductive strategy I advocate by examining the well-known argument for the existence of the planet Neptune, which is often cited as a classic example of inference to the best explanation.

Consistent with my opposition to naturalism, I do not believe it is legitimate to use facts drawn from the history of science to justify any philosophical position. Moreover, it is well-known that different interpretations of historical episodes can support various, even opposing, philosophical views. The empirical basis of historical inquiry consists of traces (artifacts such as texts, tombs, etc.), which differ fundamentally from the empirical basis provided by observation and experimentation in scientific contexts. While empirical data in science are also subject to interpretation, they consist of facts that, at least in principle, can be repeatedly tested or observed. In contrast, historical facts are only accessible through these traces and cannot be reproduced at will.

Despite the differences between observational or experimental facts and historical facts, it can still be instructive to examine key episodes from the history of science that illustrate the position I defend. In doing so, we can assess whether the four OCIM conditions mentioned earlier are satisfied by prominent scientific arguments supporting the existence of certain detected entities.

At the beginning of the 19th century, astronomer Alexis Bouvard detected (with the telescope ...) that the positions of Uranus did not conform to the predictions of classical mechanics. Several hypotheses were proposed to explain these anomalies: the influence of known planets or a comet, the presence of magnetic forces, an unknown planet, or even a revision of the mathematical formulation of the gravitational force. However, calculations based on Newton's laws suggested that the best explanation for Uranus' detected anomalies with respect to its predicted trajectory was the gravitational influence of a previously unobserved planet. This abductive reasoning led John Couch Adams and Urbain Le Verrier to conclude that an unknown planet was causing the discrepancies in Uranus' orbit.

In 1846, Johann Galle discovered this new planet, which was named Neptune, near the predicted location. (A similar reasoning process was later applied by Le Verrier to the anomalous precession of Mercury's perihelion, leading to the conjecture of a hypothetical planet "Vulcan" between Mercury and the Sun. However, Vulcan was never detected. (Baum and Sheehan 1997) In 1915, Einstein's theory of general relativity provided a new explanation for Mercury's precession, showing that Newton's theory of gravitation fails for strong gravitational fields.)

I will now attempt to show that the strength of this abductive reasoning lies in a bottom-up argument, regardless of whether Adams and Le Verrier explicitly framed their argument this way, which is a matter of historical fact.

Planets are observationally defined as bright spots that, when seen from Earth, move periodically along the constellations of the zodiac. According to Newtonian mechanics and the classical formulation of gravitational force, planets orbit the Sun and possess properties such as mass, velocity, and acceleration. By relying on the inductively confirmed causal laws of classical mechanics, we can infer from the detected effect—the anomalies in Uranus' trajectory detected through telescopic observations—the existence of its cause, namely a new planet whose motion conforms to the theory of classical mechanics.

While the discovery of Neptune is often cited as a prime example of a top-down inference to the best explanation (Douven 2021), this reasoning can also be reconstructed as a bottom-up argument. Instead of focusing on explanationist reasoning, we can construct a sound deductive argument grounded in empirical observations and inductively verified laws as follows:

1. Facts (F): Anomalies are detected in the trajectory of Uranus.

2. Inductively confirmed causal hypothesis (H): According to Newtonian mechanics, such anomalies imply the presence of a celestial body with a specific mass orbiting the Sun along a specific trajectory, which causes these anomalies.

3. Inductively confirmed association (A): Planets, defined as bright spots moving periodically along the zodiac, possess mass and follow Newtonian mechanics.

4. Alternative hypotheses (H', H''—such as a comet or magnetic forces) are not supported by observations.

Conclusion: It is more reasonable to believe in the existence of a new planet, named "Neptune".

By reconstructing the argument for Neptune's existence in this way, we see that its form is identical to the one of the "mouse argument" discussed earlier. Just as the observations of grey hair served as evidence for the presence of a mouse, the detected anomalies in Uranus' trajectory provided evidence for the existence of Neptune. The strength of the argument for Neptune's existence does not rest on the claim that it offers the best explanation for the anomalies in Uranus' orbit. Granted, we must ensure that premise H is true. While abduction leads us to consider alternative explanations for the observed anomalies, these alternatives are not rejected because they are less lovely, but because they lack sufficient empirical support. Available competing hypotheses—such as the influence of a comet or the presence of magnetic forces—are discarded *not* due to their lesser "loveliness" as explanations, but because they don't enjoy the necessary inductive observational backing.

However, premise H can only be considered likely rather than conclusively true, as we cannot entirely rule out the possibility of some unknown cause of the anomalies, even though we have no strong reason to believe in the existence of such an unknown cause.

While abductive reasoning can be useful for generating new hypotheses, it holds only heuristic value. As I have argued, abduction is not truth tropic. To evaluate the credibility of alternative explanatory hypotheses, we must investigate whether observations warrant belief in alternative causes, such as the presence of a comet, by relying on inductively confirmed laws. This process is not abductive. If alternative hypotheses lack sufficient empirical backing, they are rightly dismissed. Thus, premise H is probably true, making the deductive argument for Neptune's existence sound.

If this reasoning is correct, there were strong grounds to believe in Neptune's existence even before its shape, color, and brightness were detected. Of course, Johann Galle's subsequent telescopic observations further strengthened this belief.

Clearly, the examples of the mouse and Neptune differ in several important respects. First, belief in the anomalies in Uranus' trajectory is based on telescope images, and this belief is justified by the inductively established reliability of telescopic observations. In the case of the mouse, however, we started from immediately observed properties—such as grey hair and the disappearance of cheese—rather than images. Additionally, multiple clues were available in the mouse scenario, while for Neptune, the only initial clue was the detected anomalies in Uranus' trajectory. This is why Johann Galle's telescopic observations were particularly crucial in dispelling any doubts about Neptune's existence.

However, what ultimately justifies our beliefs in both cases—the existence of Neptune and the presence of the mouse—is the prior empirical confirmation of the relevant causal connections. These confirmations form the basis for the soundness of bottom-up deductive arguments.

To conclude, let us briefly verify that the argument for Neptune's existence meets the OCIM conditions and the requirement R. First, all the properties involved are observable, in the broad sense defined earlier. Second, the anomalies in Uranus' trajectory were repeatedly observed using a reliable telescope, yielding concordant measurement results, thus satisfying both the measurement and invariance conditions. Finally, by combining these observations with Newton's laws, a causal connection was established between the novel facts (the anomalies) and the presence of a new planet—Neptune—characterized by a specific mass and trajectory.

References

Baum, R., and W. Sheehan (1997), *In Search of Planet Vulcan. The Ghost in Newton's Universe*. Cambridge, Massachusetts: Basic Books.

De Regt, H. (2017), *Understanding Scientific Understanding*. Oxford: Oxford University Press.

Douven, I. (2021), "Abduction" in *The Stanford Encyclopedia of Philosophy* (Summer 2021 edition), ed. E. Zalta.

Ghins, M. (2024), *Scientific Realism and Laws of Nature. A Metaphysics of Causal Powers*. Synthese Library 483. Cham: Springer.

Kitcher, P. (2001), "Real Realism: The Galilean Strategy". *The Philosophical Review* 110: 151–197.

Lipton, P. (2004), *Inference to the Best Explanation*. 2nd ed. London: Routledge.

Mill, J.S. (1843), *A System of Logic: Ratiocinative and Inductive*. London: Parker.

Musgrave, A. (2017), "Strict Empiricism Versus Explanation in Science" in Varieties of Scientific Realism ed. E. Agazzi, 71–93. Cham: Springer.

Putnam, H. (1978), *Meaning and the Moral Sciences*. London: Routledge.

Quine, W. V. O. (1976), "The Scope and Language of Science", *British Journal for the Philosophy of Science* 1954, 8: 1–17. Reprinted in Quine (1976), Ways of Paradox, second enlarged edition. Cambridge: Harvard University Press.

Stanford, K.P. (2006), *Exceeding Our Grasp: Science, History, and the Problem of Unconceived Alternatives*. Oxford: Oxford University Press.

van Fraassen, B. (1995), "Against Naturalised Epistemology". In *On Quine: New essays*, ed. P. Leonardi and M. Santambrogio, 68–88. Cambridge: Cambridge University Press.

The concept of creativity in the sciences
Reflections on some problems

Hans-Peter Grosshans

Seminar für systematische Theologie, Evangelisch-Theologische Fakultät, Westfälische Wilhelms-Universität Münster, Universitätsstraße 13–17, 48143 Münster, Germany

Creativity is a much sought-after skill in many areas of society. In business, creative people are needed in product design, marketing and advertising. On the other hand, creativity is not so much in demand in the field of accounting and financial management, where solid accuracy is required—unless the finances look very bad. In football, creative players are needed in the attacking midfield, while in defence it is quite uncreative to obstruct and disrupt the play of others. The broad field of culture is very much the social field that is about creativity: in music, visual arts, theatre, dance, etc.

Lorenzo Magnani referred in his recent book "Discoverability. The Urgent Need of an Ecology of Human Creativity" even to Pope Francis, who repeatedly underlined the great importance of human creativity in his Encyclical *Laudato Si*, published in 2015. "Creativity is seen as a fundamental tool that is necessary to remedy the failures of our societies and our lives," summarises Lorenzo Magnani.[1] Following in the footsteps of Pope Francis, Lorenzo Magnani also believes that creativity is essential to solving the great problems of our time. He writes: "I am convinced that without creativity and human skills, all the other ecologies envisaged and invoked, and sustainability in general, will sadly fail".[2] Consequently, he includes human creativity among the deontological commitments: Be creative!

There is no doubt that creativity is often needed in life. But is this also the case in science?

I often see too much creativity in some sciences, such as social sciences or educational studies. There, new concepts and theories about social realities and about the education of young people are constantly being creatively designed, causing a lot of confusion in society and in education. So, the question arises: where is creativity necessary in the sciences, and where is it not? And what kind of creativity is useful in science?

In this paper, I will first follow some approaches to defining the concept of creativity in more detail, also considering its use in my field, philosophy of religion and theology. Then, in a shorter second part, I will explore the question of where and how creativity is necessary in the sciences.

---

[1] Lorenzo Magnani, Discoverability. The Urgent Need of an Ecology of Human Creativity, Cham (Springer Nature) 2022, X.

[2] Ibid.

1 What is creativity?

1.1 Creativity in psychological research

Of all the scientific disciplines, it is probably psychology that has been most concerned with creativity in recent decades. From a psychological point of view, creativity is the ability to find novel solutions to open problems or to previous solutions to problems. In this approach, creativity can be understood as an "at least six-digit relationship [...]: the action H of the individual I leading to the product P within the framework R is classified as creative by the evaluator B with regard to a system S of expectations and purposes".[3]

Products or actions are then considered creative "which are new with respect to the system of expectations of the group evaluating them and which modify this system of expectations".[4] On the one hand, creativity is understood here as a personality trait (in differential psychology), on the other hand, psychology studies the cognitive or mental processes involved in creativity (in general psychology). Four stages can easily be distinguished: preparation (gathering information), incubation (mental processing), illumination (insight), verification (checking the solution). Methodologically, creativity research in psychology is mainly conducted through biographical studies and test-oriented investigations. E. P. Torrance developed such tests (named after him) in his book "Torrance Tests of Creative Thinking" (1974). In contrast to the convergent thinking of a classical intelligence test, his tests aim to measure a person's so-called divergent thinking.

In psychological terms, the subjective dispositions of creativity include "perceptiveness, curiosity, relative independence from expectations, and imagination".[5] In the system of expectations and demands, creative people—under the condition of creative freedom—"respond with unusual and further ideas".[6] This creativity is necessary for self-preservation and for solving problems (which Pope Francis also emphasises), but it also goes beyond this and detaches itself from this practical function, as, e.g., in art.

In psychological research, the theme of "creativity" brings together different subjects and areas of research, such as research into "talent, originality, imagination, intuition, inspiration, scientific-technical invention, artistic cre-

---

[3]Wolfhart Matthäus (1976): Kreativität, in: J. Ritter/K. Gründer (Hg.): Historisches Wörterbuch der Philosophie, Basel: Schwabe Verlag. DOI: 10.24894/HWPh.2099; first published: 1976 (my translation).

[4]Ibid (my translation).

[5]Philipp Stoellger, Art. Kreativität I. Religionsphilosophisch, RGG4, vol. 4, 1738f., 1739; English edition: Philipp Stoellger, Gerhard Marcel Martin, and Josef Lukas, Art. Creativity, in: *Religion Past and Present*. Consulted online on 20 September 2022, DOI: 10.1163/1877-5888_rpp_COM_12283; first published online: 2011.

[6]Ibid.

ation".⁷ Especially with regard to the last topic, artistic creation, creativity research has a long tradition in psychology—in the evaluation of biographies of famous people or in genius research. For a long time, the aim was to analyse and understand the creative achievements of exceptional people, i.e., those that went far beyond the normal distribution of characteristics such as intelligence, diligence or work capacity. Studies on the relationship between intelligence and creativity are also part of this. Getzels and Jackson tried to prove that creativity is independent of the intelligence quotient of people.⁸ Other psychological studies seem to confirm this: Creativity and intelligence are independent.

I would like to highlight this finding from psychological studies of creativity: Creativity does not seem to be identical with superior intelligence, but can sometimes even be the result of ignorance on the part of the intelligent. It should be added, however, that the identification of creativity is always subjective. The awareness of a creative deviation from the norm always depends on the fundamental attitudes of the person making such determinations.

In my academic field, theology and philosophy of religion, in such a situation, where the thing that is grasped with a concept like creativity threatens to become blurred—and this precisely also through subjective relativisation—one turns to the definition of the concept (i.e., creativity), which is then also ideally conceived.

1.2 Creativity in the perspective of theology and philosophy of religion

The concept of creativity plays an important role in the philosophy of religion and in theology. This is particularly true in relation to God the Creator. In Christianity it is believed that God has the creativity to create a complete world (universe) out of nothing. It is understood as pure free creativity to begin a new state without reference to anything that already exists. Plato's Demiurge, who orders the existing chaotic matter into a world, a universe, in analogy to his own perfection, or Aristotle's Unmoved Mover, who orders everything as an unsurpassable state of perfection, i.e., towards himself as the ultimate goal, are also creative in a certain way. That their creativity is limited becomes clear in relation to the Christian God, who in a free creative activity creates the world out of nothing.

According to Immanuel Kant, such creativity is synonymous with freedom par excellence. As Kant defines it: Freedom is "the ability to bring about a state of affairs of one's own accord, the causality of which, according to the law of nature, is not in turn subject to any other cause which determines it

---

[7]Wolfgang Matthäus, Art. Kreativität, fn. 3 (my translation).
[8]Jacob W. Getzels and Philip W. Jackson, Creativity and Intelligence, New York 1962.

according to time".[9] Kant called this a pure transcendental idea—the idea of something outside experience, a kind of limiting concept of all experience. If we take the idea of creativity here out of its specifically Christian background and formulate it more generally in terms of the philosophy of religion, we can profile it by its opposite. "From the point of view of the philosophy of religion, creativity shows various accents in response to its opposites: creativity stands for innovation in contrast to tradition; it is the 'new' in contrast to the 'old'; it stands for the transcendence of mimesis in contrast to nature. Despite all the correlations, creativity cannot be extrapolated from what has gone before".[10]

We find Kant's idea of freedom, which we have just mentioned, originally elaborated in the concept of creativity in Christian theology. This is the idea of the creation of the world (in the sense of the universe) out of nothing. In the doctrines of theology, the concept of creativity is related to God and thus defined in an ideal way. In principle, theological doctrines always proceed in an idealising way—i.e., they do not attempt to determine the objects they deal with in a generalising way on the basis of a lot of data, nor do they attempt to do so experimentally.

A concept acquired in relation to God, such as the concept of creativity, then functions in relation to the less ideal conditions of the empirical world as a borderline concept to which any talk of creativity must ultimately be oriented.

With the assumption of the creation of the world out of nothing, a purely free creativity is attributed to God in order to start a new state without referring to anything that already exists.

In the course of its history, Christian theology soon developed an understanding of nothingness as pure nothing: it was thus understood—in the Greek vocabulary—not as μὴ ὄν but as οὐκ ὄν. To speak of creation from nothing (*creatio ex nihilo*) is to think of nothing. But nothing taken for itself cannot be thought. We can only think nothingness in distinction from what exists.

What is meant by nothingness can then be thought in at least two ways. On the one hand, nothingness can mean the radical opposite of being. In Greek this is expressed by the negation οὐκ ὄν—nothingness as the complete opposite of being.

On the other hand, nothingness can mean the negation of something that exists—that is, what it no longer is or has not yet become. In Greek this is expressed by the negation μὴ ὄν. Nothingness is understood in this sense as a not-yet, which carries with it various possibilities that are not (or not yet)

---

[9] Immanuel Kant, Kritik der reinen Vernunft, B 561.
[10] Philipp Stoellger, Gerhard Marcel Martin and Josef Lukas, Josef, Art. Creativity, in: *Religion Past and Present*. Consulted online on 20 September 2022, DOI: 10.1163/1877-5888_rpp_COM_12283; first published online: 2011.

realised. If I were to say, as Plato did, that God created the world out of the matter that was present to him, then I would understand "nothing" in this negating sense: it is not (yet) this and that—whereas it could be anything.

When the Christian faith speaks of creation out of nothing, by contrast, nothingness is understood as a complete, fundamental non-being. Such nothingness cannot necessarily be defined. It can be addressed (e.g., in myths), but not thought of in a strict sense.

The attention of theology, in its statement of a creation out of nothing (*creatio ex nihilo*), is then less on the "nothing" than on the Creator. The logic of action, as it were, dominates the understanding of creativity. The emphasis is thus on the fact that in creating out of nothing, the Triune God affirms Himself, and thus His Godhead, and is creative by connecting Himself to nothing other than His own Being.

The "nothing" here is understood as pure nothingness (οὐκ ὄν)—and not as not-yet-being in the sense of not-yet-realised possibilities that are being explored anew in reality, which contains many possibilities. Martin Luther put it very succinctly: "*creare est semper novum facere*: creativity means always creating something new".[11]

So it is said of God that he creates something out of nothing. Not only is something new created, but a completely new beginning is made by God. Creativity, in the strictest sense, is an act of innovation par excellence. God creates something new—something new in relation to himself. What is "new" about it?

The newly created is an Other in relation to God—an Other in which one's own self is not actually realised, but in which one's own self is transcended.

To quote Martin Luther once again: "*haec est natura Dei, [...] ex nihilo creare omnia*: it is God's nature to create everything out of nothing".[12]

In the evaluative perspective of the Christian faith, creation out of nothing expresses that it is love out of which God affirms others than himself and thus allows nothing to become something.

If we generalise this consideration with regard to human creativity, the hallmark of creativity would not only be an innovative, new solution to a problem, but also the transgression of the human being out of himself towards an Other, which, strictly speaking, is actually a stranger. A creative person has to leave—transcend—himself, that is, his previous understanding of things and the familiar ways—and not just vary what is known so far.

God is a subject who begins with himself and only begins with himself. In the model of the divine Trinity, in which some theologians systematised the diversity of Christian talk about God and the diversity of human experience

---

[11]Martin Luther, Resolutiones disputationum de indulgentiarum virtute. 1518, WA 1, 563,8.
[12]Martin Luther, WA 40/III, 154,11f; cf. WA 1, 183,39f.

of God, this is expressed in the figurative but actually modelling talk about God the Father. In fact, the model of the divine Trinity is a product of theological creativity. As Father, God is par excellence the one who begins, the one who acts on his own initiative. In the Trinitarian model, God the Father stands for the fact that God is pure activity and inexhaustible possibility in one. This is nothing less than absolute freedom. As Father, God is the free and therefore the creative par excellence.

Whereas we, in the context of our world, can only begin in such a way that something is already given, God the Father can make an absolute beginning—that is, a beginning without any specification; a creative beginning out of nothing.

While for us the rule is: nothing comes from nothing, for God it is true that he can make everything out of nothing.

Such ontological creativity only became conceivable "in the expansion of the horizon of possibility, when the concept of infinity was transferred to God [...] with a 'turn' from the primacy of reality to that of possibility".[13] This can already be seen in Gregory of Nyssa and is then developed, above all, by Duns Scotus and William of Ockham.

In the early modern period, at the latest, the idea of the perfection of nature dissolved into an awareness of a contingent constellation that "no longer had to be imitated, but artistically and technically transcended".[14] Georg Wilhelm Leibniz took this broadening of horizons even further. For a long time, only God's creativity was regarded as essential and human creativity as non-essential (since it was ultimately understood only as imitation [*mimesis*]). When the latter became independent in art, but also in technology and science, the legitimacy of human creativity could no longer be denied.

At the same time, however, its ambiguity became apparent. From a theological perspective, human creativity is always seen as ambiguous. It becomes particularly problematic when it is equated and confused with God's creativity. In any case, the creative energy of the human spirit is always judged critically, despite the appreciation of successful creations. From a theological perspective, the results of human creativity are always ambiguous and to be used in different ways. However, this critical view of human creativity is itself misguided if it defames human creativity.

The capacity for creativity is theologically given to human beings by being created in the image of God, and it is part of human beings' God-given freedom to be creative themselves. Therefore, a non-use of creativity and an inertia of the spirit would also be contrary to this gift of God. This is especially true with regard to the creativity made possible by the gift of

---
[13] Philipp Stoellger, Art. Creativity, fn. 9.
[14] Ibid.

language, through which man can also deal with problems by inventing new metaphors and models and thus open up new perspectives.

In order to structure the reflections on the theological concept of creativity systematically, we can specify the analysis of creativity in theology and philosophy of religion in at least four ways: according to a logic of origins, a logic of action, a structural logic and a semiotic logic.

If we look at creativity according to the logic of origins, creativity belongs "to the essence of all that is"—which in the world of religions leads us to myth, in which this origin and creativity is expressed narratively.

If we look at creativity from the logic of action, we must conclude that creativity is an action that presupposes a creative subject. It is only since the Renaissance that this has also been attributed to human beings—whether in competition with God's creativity or as a consequence of human beings being made in the image of God. Interestingly, human creativity has been more often theologically criticised than positively appreciated. But philosophically, too, the emphatically creative human subject has been criticised, especially when it has been linked to the cult of genius.

Looking at creativity in terms of structural logic, creativity is understood "as the purpose of a process in which something new is created. The advantage (but also the limitation) of these perspectives is that they do not rely on a creative subject. Moreover, attention is drawn to the emergence of ontological innovation [...]. Signs are not only interpretations, but also existing relations".[15]

Finally, creativity can be explained semiotically "as a quality inherent to a subject or process, in that both the process of signification and the user of signs are capable of being creative". Here we can refer to Charles Sanders Peirce's concept of "abduction". "Signs are not only interpretations but also stand for real relations. The use of signs is therefore an act by which real things are created"—as in speech acts. "In relation to a natural language, the specific manifestations of creativity are metaphors and allegories".[16]

2 Creativity in the sciences

Human creativity cannot create something out of nothing. Under the conditions of finiteness, however, a *"creatio ex aliquo"*, a creation out of something, is possible. "Creative processes do not start from nothing, they do not begin with nothing, but they presuppose something, they are connected to something that they transform, reform or even radically revolutionise. [...] But the something to which creativity is linked cannot be determined and constructed on the basis of predetermined rules. [...] If creativity and what is designed, built, [...] formed with the claim of being creative could

---
[15] Ibid.
[16] Ibid.

be reduced to a rule-based, learnable and reproducible knowledge, it would be superfluous to speak of creativity as the creation of something new. [...] In this respect, creativity is more and different than mere technical problem solving. By referring to something that has gone before, it refers to unfinished, inexhaustible possibilities. And sometimes it is necessary to think the impossible in order to make the possible real and to sharpen the sense of the possible. [...] Thinking creatively and promoting creativity in scientific, artistic and organisational contexts would therefore mean giving space and time to thinking in the subjunctive and in the potentialis".[17]

This means that creativity always includes the possibility of failure and lack of success. In science in particular, the failure or non-success of scientific projects is not seen as a good thing. If experiments do not produce useful results, they are usually not published, even though they may provide insight. And also: Many new hypotheses or theories have met with incomprehension among contemporaries. Failure can therefore also mean that the time was not yet ripe for a new insight or observation. But even an error, which in a way is also a failure of a scientific investigation, is not meaningless on the way to truth. You only get to the truth if you risk the error. Nor can efficiency be a criterion of success in science. Researchers who pursue their curiosity are often inefficient, ponderers, collectors or tinkerers obsessed with a question, a problem or an idea. Eagleman and Brandt write: "Creative output typically requires many failed attempts. As a result, across human history, new ideas take root in environments where failure is tolerated. [...] Human culture is littered with ideas that have been rejected by the public and passed into oblivion."[18]

At the beginning of this paper, I raised the question of where creativity is needed in science and where it is not. I added the question of what kind of creativity is useful in science. I have just given an initial answer to these questions. It makes sense for new research questions, approaches and methods to be used in the sciences. "New" here means genuinely innovative, leaving behind previously trodden paths, not just varying previous research. For young researchers, this is very risky, because the result of such research can also be that the research question is unproductive, the research approach does not lead to results, or the method is not really helpful. This usually means that a scientific career ends before it begins. In this respect, it is more important for established researchers to be creative in this sense and to go beyond their own successful research profile to really pursue new questions and develop new methods.

---

[17]Andreas Großmann, Kreativität als Denken und Praxis des Möglichen. Zur Einführung, in: Andreas Großmann, Kreativität denken, Tübingen 2020, 1–7, 3f. (my translation).

[18]David Eagleman and Anthony Brandt, The Runaway Species: How Human Creativity Remakes the World, Edinburgh 2017, 176, 180.

A second response: With regard to the methodisation of new research approaches, the abductive method is particularly interesting in terms of creativity in science.[19] Charles Sanders Peirce reintroduced this method into the philosophy of science. For Peirce, scientific statements are about a context of discovery, not a context of justification (as with the positivists). According to Peirce, it is the logic of discovery that is important, not the logic of inquiry. "Abduction is that kind of argument which proceeds from a surprising experience, that is, from an experience contrary to an active or passive belief. This takes the form of a perceptual judgement or a proposition referring to such a judgement, and a new form of belief becomes necessary to generalise the experience". "Deduction proves that something must be; induction shows that something is actually operative; abduction merely suggests that something might be."[20]

Thus, an explanatory hypothesis is formed abductively on the basis of an experience or observation that surprises or irritates, disturbs or challenges the usual experiences and observations. Predictions are derived from the hypothesis and facts are sought to verify the hypothesis. Abduction is thus the starting point of the actual cognitive process that follows perception.

Peirce also emphasised the creative moment and the originality of the idea that appears like a flash of lightning, while being aware of the ambivalence of this spontaneous event. "The abductive conjecture comes to us in a flash. It is an act of insight, albeit an extraordinarily deceptive insight. It is true that the various elements of the hypothesis were previously in our minds; but the idea of bringing together what we had never dreamed of bringing together before, flashes the new conjecture into our contemplation".[21]

As a theologian, I see an analogy here with a mystical insight, the point of which is also that after a long contemplative preparation—an exercise in concentration—the thing to be known suddenly falls into thought. This is the beginning of thinking, of reflecting on the thing that has thus become present, and thus the actual process of cognition.

Analogies can also be found in Aristotle, for whom philosophy begins with astonishment, surprise and wonder. He had in mind above all the irritating astonishment, e.g., that something is not in its usual place, which triggers thinking and makes one ask about the preconditions that led to the irritation, but also about the preconditions of the usual order.

In Christian theology, too, amazement (surprise and wonder) has a cognitive meaning. But while philosophy seeks to overcome the surprise and wonder that go hand in hand with ignorance, amazement remains

---

[19] Cf. Lorenzo Magnani, The Abductive Structure of Scientific Creativity. An Essay on the Ecology of Cognition, Cham (Springer Nature) 2017.
[20] Charles Sanders Peirce, Collected Papers, Vol. 5, ed. by Charles Hartshorne, Cambridge 1934, 171.
[21] Ibid., 181.

permanently present in faith. There, too, people are astonished because they hear and see something. Here, too, amazement is provoked by something unknown and leads to knowledge. But this unfamiliarity does not become ordinary and normal. Rather, the following applies: "The more you recognise the amazing, the more amazing it becomes".[22] This, of course, has to do with the mysterious character of faith and thus with the amazing God. For this reason, theological knowledge does not come to a standstill in the end, but is always set in motion anew by this amazement. In view of this, I would conclude that creativity in science presupposes a sense of amazement, a sense of surprise and wonder—be it frightening or joyful—and thus being intellectually moved and taken in.

A third answer to the question of where creativity is needed in the sciences, and what kind of creativity is useful, can be given in relation to the evaluation of observations, experiences and data. Semiotically, creativity is needed to find a new language for new knowledge. This is particularly evident in the creation of new metaphors and new models.

My subject, theology, offers many examples of this. They can already be found in the Hebrew Bible, where the understanding of God was creatively developed again and again with new metaphors and models. Such examples can then be found especially in the New Testament, already in Jesus and the evangelists who portrayed him, and then even more conceptually in Paul. Paul, in particular, had to recognise that the old metaphors, images, narratives and models could only inadequately capture and express what happened with the incarnation of the Logos in Jesus Christ. New metaphors, images, narratives and models had to be formulated and accompanied by a new religious practice. You cannot put new wine in old wineskins, but in new wineskins (Mt 9:17).

This creative new way of thinking about God, documented in the New Testament, also required new models of God. The doctrine of the Trinity of God is basically a simple, complex model of God in which the many different ways of talking about God in the New Testament have been brought together and thus been modelled in the contexts of life with a variety of experiences of God. The Trinitarian model of God is an excellent example of creative intellectual innovation because it allows for a whole new understanding of God (for example, that God's unity does not exclude plurality). At the same time, this model works back to the interpretation of the diversity of life in the horizon of God. The Trinitarian model of God has a hermeneutical function in relation to the human situation. For it makes our own life situation understandable to us in such a way "that we find ourselves created by God and fallen with him, that as such we are found by God through Jesus Christ, and that we are guided by the Holy Spirit to find the right way to the goal

---

[22]Eberhard Jüngel, Zum Staunen geboren. Predigten 6, Stuttgart 2004, 10.

and end of life".[23] The trinitarian model is complex because it not only models a diversity of discourses of God in the Bible, but also preserves rather than harmonises the diversity of God's relation to human beings. A new creative innovation in thinking about God would have to at least match the capacity of the Trinitarian model. Anything else would be neither creative nor innovative.

From this I would like to draw a general conclusion for the topic of "creativity in science": Creativity in science not only cannot be prescribed, it cannot even be demanded. If creativity is lacking in science, this is not a fundamental deficiency. Uncreative science can be very sound. We cannot, for the sake of research creativity, constantly replace models and theories that have proved their worth with new ones—and certainly not if they do less than the previous ones.

Nevertheless, science must remain open to new creative models, metaphors, theories and the like—for the sake of a better understanding of the problems and issues being researched.

---

[23] Gerhard Ebeling, Dogmatik des christlichen Glaubens, Vol. 1, Tübingen 1979, 545.

Justification, creativity, and discoverability in mathematics: The example of predicativity

Gerhard Heinzmann

Archives Henri-Poincaré (UMR 7117), Université de Lorraine, Université de Strasbourg, CNRS AHP-PReST, F–54000 Nancy, France

> **Abstract.** The main thesis put forward in this paper is that the norm of predicative definability is a relative concept which therefore greatly affects its philosophical relevance. Predicative definitions even risk being a miss if they are not considered as an indication to abandon classical continuous analysis.
>
> After an overview of the developmental stages of the mathematical specification of the intuitive concept of predicativity as per Russell and Poincaré, in section three I discuss different historical approaches to justify the principle of complete induction, and ask whether it is possible to avoid its impredicativity. If predicativity is considered from an extensional perspective, complete induction would possess an irreducible impredicative character even though it is not treated as an explicit definition but as an inductive definition. By contrast, if predicativity is considered from an intensional perspective, a purely operational and predicative justification of complete induction using operative imagination (Lorenzen) would be possible.
>
> Mathematical practice depends on epistemic norms, which are themselves influenced by mathematical developments which, in turn, influence the standards for ontological, epistemological and semantical questions.

1 Introduction

There is no standard account of the condition to be fulfilled in order to justify the general concept of mathematical definition, nor how the distinction between suspicious procedures should be drawn. In efforts to justify the epistemically significant nature of mathematical definition, different criteria have been proposed, some of them depending on the framework of object-realism, formalism, and intuitionism. I will confine myself to the criterion of predicativity, which was advanced as philosophically motivated by proponents of a variant of constructivism.

My aim is to show that even with respect to predicativity there is only a vague context-dependent boundary between evident and suspect definition: the mathematical implementation of the normative philosophical criterion remains vague both in terms of its technical scope and its philosophical stringency.

In the second section, I take Russell's and Poincaré's perspective as a starting point in order to present the technical specification of the concept of predicativity in terms of its historical development. In the third section, two attempts to justify the induction principle by Hilbert/Bernays and Lorenzen are then examined as a case study with regard to its predictive character.

2 Predicative definability: from Russell-Poincaré to Weyl, Lorenzen and Wang

For a detailed version of this section, see Heinzmann & van Atten (2022, 223–256).

In the wake of the discovery of the famous Russell paradox, Bertrand Russell and Henri Poincaré set out to answer the question 'Which propositional functions define sets in a non-circular way?' In a second step, the creativity of the search centered on the more general question 'Which mathematical sets (respectively concepts) are non-circularly definable?'.

In his *Principles of Mathematics*, Russell noted:

> "Having dropped the former [the axiom of comprehension], the question arises: Which propositional functions define classes which are single terms as well as many, and which do not? And with this question, our real difficulties begin" (Russell 1903, 103).

In 1907, he introduced a new terminology to solve the problem:

> "Norms[1] (containing one variable) which do not define classes I propose to call *non-predicative*; those which do define classes I shall call *predicative*" (Russell 1907, 34).

In his discussion of Russell (1907), Poincaré (1906) confirmed the non-predicative direct or indirect definitions (existence postulates = propositions) as circular: he labeled a definition 'predicative' if, in the *definiens*, the *definiendum* does not occur, and no reference is made to it: Otherwise, it was considered non-predicative.

For Poincaré, the circularity based on a non-predicative definition in Russell's antinomy is the sign of the Cantorian's "realistic" error, i.e., considering a totality as a *datum* independent of the construction of its individuals.[2] This is trivially the case for actual infinite sets that are to refuse.

Poincaré then gave two definitions of predicativity:[3]

$P(1)$ leads to the idea of a predicative definition which imposes a limit on the unrestricted quantification over sets which are available to us in a "constructive" sense: *it places a constructive restriction at the object level.*

$P(2)$ indicates restrictive conditions imposed on the quantification without an explicit restriction of the domain: for a classification to be predicative, it is sufficient that the quantification over an indefinite domain,[4]

---

[1] Russell calls propositional functions "norms" here.
[2] The definition of E in $\forall X(X \in E \leftrightarrow X \notin X)$ is non-predicative, since the *definiendum* E is itself a possible totality of the variation domain of a universal quantifier.
[3] Cf. Heinzmann (1985, chap. IV).
[4] A domain is indefinite, if we can add an element to it which cannot be expressed by the means of previously fixed definitions.

on which the *definiendum* depends, does not change the already determined classification of its elements: *it places a constructive limitation at the level of description.*

There remains the problem of the extensional equivalence between $P(1)$ and $P(2)$. Will they exclude the same definitions? As we will see, new light was shed on this question only in the 1960s.

In his seminal book *Das Kontinuum* (1918), Hermann Weyl held that Russell's way out of his antinomy, the "type theory", made mathematics unworkable. Weyl is in all probability the initial proponent of the predicative definition of real numbers. It is based on an *iterative* formation of ideal objects with respect to the domain of natural numbers, equipped with its operations and presupposed—contrary to Poincaré—in a Platonist way. Weyl rejects the set-theoretic reconstruction of natural numbers, as our grasp of the basic concepts of set theory depends on a prior intuition of natural numbers.

In his system, he called a formula arithmetical if it does not contain bound *set* variables. An arithmetical formula defines a property that refers only to the totality of natural numbers but does not refer to the totality of sets of natural numbers. This leads to a system, called **ACA**, containing an Arithmetical Comprehension Axiom:

$$\exists X \forall x [x \in X \leftrightarrow \varphi(x)]$$

for each arithmetical φ, where the variable X is not in φ.

The system **ACA** is a conservative extension of Peano Arithmetic, even though it employs second-order concepts. This enables Weyl to recover a substantial amount of Analysis. Nevertheless, predicative mathematics is restricted to arithmetically definable sets. What matters is that the usual *Least Upper Bound Axiom* (**LUB**), which states that every set of reals which is bounded above has a least upper bound, is not valid because it involves an impredicative definition (Weyl 1918, 77).

So, the question arises: Can the technical implementation of predicativity lead to a less constrained revision of mathematics?

This brings us to the work of Paul Lorenzen. According to him, the requirement to accept only "definite" propositions is a methodological boundary to fully grasping the aspect of mathematics which can be regarded as "stable" or "safe". He defines 'definite' as follows:

(1) Any proposition decidable by schematic operations is called "definite".

(2) If a definite proof or refutation concept is fixed for a proposition, then the proposition itself is also definite, more precisely proof-definite or refutation-definite" (Lorenzen 1955, 5–6).

He notes that

(i) Non-predicative *concept formation* (sic) is indefinite and therefore excluded from operative mathematics.

(ii) Quantifiers are permissible provided that the formulas in the quantification domain are definite.

(iii) The natural numbers and Peano axioms, together with the definitions of addition, multiplication, and exponentiation, can be constructed as definite.

However, Lorenzen's real numbers are not a model of an ordered and complete Archimedean field.

Hao Wang further developed the idea of predicative mathematics as a justified part of mathematics with explicit reference to Lorenzen. Both, Wang and Lorenzen, aimed at transfinitely iterating the construction of definable sets in their systems of ramified analysis. Nevertheless, unlike Lorenzen, Wang accepted classical logic.

Wang's idea was to start from a multi-layered constructive set-theoretical ordered hierarchy and to ask whether one can then provide a more accurate characterization of predicativity (Wang, 1964, 578). His most important contribution was the discussion of the relationship between predicativity and ordinals. He related predicative defined sets to constructive ordinals by establishing a hierarchy as the union of all systems Σ_α, where α is a constructive ordinal.

This hierarchy does not lead beyond recursive ordinals! (Spector 1955). Kreisel subsequently formulated the thesis that all predicatively definable sets belong to Σ_{ω_1}, where ω_1 is the upper bound of recursive ordinals. Wang's hierarchy is a prime example of formalizing the intuitive idea of predicativity expressed by Poincaré in his first definition $P(1)$: it limits the quantification on already constructed sets, which comes down to a restriction in terms of construction.

Assuming the totality of natural numbers,[5] and presupposing a 2$^\text{nd}$ order language enabling quantifications over sets of natural numbers, on the basis of preliminary work by Georg Kreisel (1960), Solomon Feferman (1964) proposed two definitions to predicativity: one which amounts to a constructive restriction at the object level and corresponds to $P(1)$, and one which comprises an extension of the domain of their 2$^\text{nd}$ order quantifiers and corresponds to $P(2)$. He then shows that, in both cases, the predicatively definable sets are the same. In this way, Poincaré's intuition is confirmed at a higher level.

---

[5]The question of their predicativity is not addressed.

Finally, in 1959 Stephen Cole Kleene proved that Σ_{ω_1} exactly coincides with what is referred to as the class of hyperarithmetical sets Δ_1^1. Does Δ_1^1 express the central idea of predicativity in a clear way? No, because we can conclude from the falsity of a proposition, if relativized to Δ_1^1, to its non-predicativity, but not from the validity of a theorem in hyperarithmetical analysis to its predicativity. Predicative analysis seems to be somewhere between arithmetic and hyperarithmetical Analysis.[6]

There is yet another difficulty. The class of hyperarithmetic sets only specifies for the non-predicativiste what the predictive universe should be: Indeed, for a recursive ordinal it must be proven that not only its definition is predicative, but also that its ordering is predicatively recognized as being a well order by using principles of reasoning that had already been shown to be predicatively acceptable at a previous stage (Kreisel 1960, 387). This is why Kreisel, Feferman, and Kurt Schütte introduced the concept of *predicative provability*. I am unable to discuss this concept here.[7]

3 Is it possible to avoid the impredicativity of the induction principle in mathematics?

According to Hilbert's and Bernays' *Grundlagen der Mathematik* (1934, §2), the method of proof of complete induction is obtained from iteration by a further step involving "experiments in the mind". How should we thus imagine this further step of an experiment in the mind? In fact, it is obtained by adding to the iteration schema

(a) $\Rightarrow$ I ["we can construct I"]

(b) $n \Rightarrow n$I ["if we have n, we can construct nI"]

(S)

the final clause

(c) We can obtain all numerals by application of the scheme S.

This clause (c) does not follow analytically from clauses (a) and (b) of S.

---

[6] Starting from arithmetical sets or relations, we obtain Σ_1^1 and Π_1^1 by an existential (respectively universal) quantification of the second order. Δ_1^1 designates the intersection of these two classes.

[7] In the 1960s, independently of each other Kurt Schütte and Solomon Feferman discovered that a certain ordinal limit Γ_0 plays for the so-called predicative Analysis a role analogous to the one played by ε_0 for arithmetic: for each well order of type $\alpha < \Gamma_0$ there is a proof that the order in question is a well order and that it uses exclusively order types $< \alpha$, i.e., it is the smallest ordinal whose well order is no longer *predictively* provable.

Modulo some further abstractions, we have thus returned to what is called a recursive definition of natural integers:

(a) $N(\mathsf{I})$ ["I is a number"]

(b) $N(n) \Rightarrow N(n\mathsf{I})$ ["if n is a number, its successor is also a number"] \hfill (S′)

(c) We can obtain all numbers by application of (a) and (b).

Now, the justification of the *schema* of complete induction

$$[E(0) \wedge \forall x(N(x) \wedge E(x) \to E(x'))] \to \forall x(N(x) \to E(x)) \qquad (\text{T})$$

where T applies to any property E is correlative to S′, which means that the final clause c) cannot be deduced from the clauses a) and b): an application of T is needed. In other words, T implies S′, and S′ implies T. However, the final clause expressing that we can obtain all numbers by application of a) and b) amounts to taking N to be 'minimal', but N should be first and foremost defined by (a), (b), and (c)!

As a result, even without using an explicit second-order definition for induction, this inductive definition is impredicative. Other examples of the attempt at predicative reduction of induction are discussed in Parsons (1992).

Expressed in a terminology I introduced in a recent article on thought experiments (Heinzmann 2022), numerals constructed using the rule S constitute the *experimental realm* belonging to the general Kantian scheme G (universal) of string repetition. S is imaginatively (by means of a very far-reaching intuition) related to S′, which is symbolically interdependent with T.

Hilbert/Bernays 'sees' in an apocryphal (intuitive) way the relation between the iteration rule S and the inductive definition S′ without being able to deduce S′ by logical means: *it is a genuine thought experiment* that could be confirmed by examples of real experiments (=calculations).

In mathematical thought experiments, we take recourse—based on mathematical experiments—to a modal deviation using further semiotic means. *Epistemic intuition* provides us with access to these deviations as 'genuine possibilities' of mathematical inferences as opposed to 'pure fictions'. This accessibility is the justification for the validity claim concerning mathematical thought experiments (Heinzmann, 2022).

Now, Lorenzen argues that in his operative approach it is possible to obtain the Peano axioms by avoiding the impredicativity of an inductive definition of induction without recourse to an apocryphal intuition. In fact, for him, the induction principle is a predicative (= definite) *meta-rule* (independent of the language level of A) of the form

$$A(I); A(m) \to A(m\mathsf{I}) \Rightarrow A(n), n \text{ arbitrary}$$

that constitutes an operative interpretation of the classical induction principle. In fact, the formula in the conclusion is definite: its range consists exclusively of numerical signs constructed according the rules

$$\Rightarrow \mathsf{I}$$
$$m \Rightarrow m\mathsf{I}$$

and all numerical signs are the result of such a construction.

The acceptance of the predicative induction rule firstly points to a shift in meaning concerning the term used: impredicativity no longer refers to the definition of sets, but of concepts! Therefore, Parsons (1992, 152–154) is correct in his assertion that Lorenzen's concept of predicativity is novel and that is not so firmly entrenched as the classical interpretation of Poincaré's or Russell's definition of predicativity. However, it is not incompatible with Poincaré's definition: the circle lies in the fact that one speaks of sets that could not be extensions of *predicates* antecedently understood. In the same way, for Lorenzen, sets always should be in the range of predicates antecedently understood as definite: we should respect the conceptual order that places the understanding of the predicate before the apprehension of its extension as an object. He eschews set theoretic realism and considers the inductive rules as giving us an understanding of the predicate 'natural number'. The understanding of the predicate occurs prior to the insight that the set exists (Parson 1992, 254).

Lorenzen's formulation of induction can thus be read as a special case of an application of a constructive version of the ω-rule, which states that given a recursive function f such that for every natural number n the value $f(n)$ is the Gödel number of a proof of $A(n)$, one may proceed to: *for all n, $A(n)$*.

One thus obtains a complete semi-formalism of arithmetic without (probably) using actual infinity. However, the scope of 'extensional' and 'intensional' predicativity is now different: Kreisel (1959) shows that the Cantor-Bendixson theorem in Analysis (every closed set is the union of a perfect set and of a countable set) involves impredicative definitions, given that it does not hold in Δ^1_1, whereas Lorenzen and Myhill (1959) implicitly ascribed predicativity to the theorem as a consequence of their use of generalized inductive definitions.

4 Conclusion

The uncertainty of the predicativity of complete induction remains open. There are two kinds of predicativism settled between realism in extension (Platonism) and intuitionism:

I. Extensional predicativism (Poincaré, Weyl),

II. Intensional predicativism (Poincaré, Lorenzen).

In the first case, predicative definability leads to thought experiments—where we resort to a modal deviation of a logical inference using further semiotic means, in the second case to a semi-formalism. The difficult question to decide is this: What is preferable for understanding complete induction, a thought experiment (Hilbert) respectively pure intuition (Poincaré) on the object level, or operative imagination on the 'practical' meta-level? If one is convinced that the first important thing in mathematics is not proof but conceptual construction, Lorenzen's predicative definiteness implying a revisionist position gives predicative insights into classical non-predicative constructions.

Is predicativity a miss or perhaps a hint that one should abandon full formalisms, or even continuous Analysis, or should one return to "pre-Cartesian" geometric-topological intuition, as suggested by Poincaré and Bernays (1979, 13–14)? In philosophical terms, continuity cannot be adequately described by a full formalism without being a set theoretical realist. Nonetheless, difficulties in such a realism were precisely the motivation for Poincaré to invent predicative definability ... and now predicativity requires thought experiments or semi-formal systems! We continue to remain in a state of vagueness.

Acknowledgements. I thank Paolo Mancosu for an exchange on predicativity which motivated this presentation. Lectures in Irvine (April 2023) and Buenos Aires (DLMPST, July 2023) were based on variants of this paper. I would like to thank the Service "traduction" de la MSH Lorraine for correcting the English language.

References

Bernays, Paul (1979). "Bemerkungen zu Lorenzens Stellungnahme in der Philosophie der Mathematik", in K. Lorenz (ed.), Konstruktionen versus Positionen. Berlin/New York: De Gruyter, vol. I, 3–16.

Feferman, Solomon (1964). "Systems of Predicative Analysis", The Journal of Symbolic Logic 29: 1–30. Heinzmann, Gerhard (1985). Entre intuition et analyse. Poincaré et le concept de prédicativité. Paris: Blanchard.

Heinzmann, Gerhard and van Atten, Mark (2022). "Révisionnisme mathématique: l'enjeu constructif", in: Précis de philosophie de la logique et des mathématiques, sous la direction de A. Arana et M. Panza. Paris: Editions de la Sorbonne, 199–256.

Hilbert, David and Bernays, Paul (1934) Grundlagen der Mathematik I. Berlin/Heidelberg (Berlin, Heidelberg, New York, Springer 1968).

Kleene, Stephen Cole (1959). "Hierarchies of Number-theoretic Predicates", Bulletin of the American Mathematical Society 61: 193–213.

Kreisel, Georg (1959). "Analysis of the Cantor-Bendixson theorem by means of the analytic hierarchy", Bulletin de l'Academie Polonaise des Sciences. Serie des Sciences Mathematiques, Astronomiques et Physiques, vol. 7: 621–626.

Kreisel, Georg (1960). "La prédicativité", Bulletin de la Société mathématique de France 88 : 371–391.

Lorenzen, Paul (1955). Einführung in die operative Logik und Mathematik. Berlin/Heidelberg/New York: Springer, (21969).

Lorenzen, Paul and Myhill, John (1959). "Constructive Definition of certain Analytic Sets of Numbers", The Journal of Symbolic Logic 24: 37–49.

Parsons, Charles (1992). "The Impredicativity of Induction" in: M. Detlefsen (ed.), Proof, Logic and Formalization. London/New York: Routledge, 139–161.

Poincaré, Henri (1906). "Les mathématiques et la logique", Revue de Métaphysique et de morale 14: 294–317.

Russell, Bertrand (1903). The Principles of Mathematics. Cambridge: Cambridge University Press.

Russell, Bertrand (1907). "On some Difficulties in the Theory of Transfinite Numbers and Order Types", Proceedings of the London Mathematical Society., 2nd Ser., Vol. 4: 29–53.

Schütte, Kurt (1960). Beweistheorie. Berlin: Springer.

Spector, Clifford (1955). "Recursive well-orderings", The Journal of Symbolic Logic 20: 151–163.

Wang, Hao (1964). A Survey of Mathematical Logic. Peking/Amsterdam: Science Press/North-Holland.

Discoverability: affordances and eco-cognitive situatedness
Towards an ecology of human creativity

Lorenzo Magnani

Department of Humanities and Computational Philosophy Laboratory, University of Pavia, Piazza Botta 6, 27100 Pavia, Italy
E-mail: `lorenzo.magnani@unipv.it`

Recent studies in the field of "EEEE" cognition (extended, embodied, embedded, and enacted) have demonstrated that the role of what I called environmental situatedness can be a useful way to understand human cognition and its evolutionary dimension. This means that rather than storing detailed representations of the environment and its variables in their memory, humans actively modify it by obtaining information and resources that are either already available, extracted from the environment, or created from scratch. In other words, resources and information are not only provided; they are actively sought after and even created. We may think of human cognition as a chance-seeker mechanism in this way. Thus, chances are not only information, they are also "affordances," that is, environmental anchors that help us make better use of outside resources. Certainly, discoverability depends on having the right affordances available. Even still, abduction is significant because it clarifies all those hypothetical conclusions[1] that are controlled by activities that consist of deft environmental manipulations to find new affordances as well as the creation of artificial external items that provide new affordances or signals.

---

[1] Inference is often understood in terms of logic or psychology. Conversely, as I shall elucidate later in this chapter, I approach the concept of inference (and hence the hypothetical abductive inference) from a Peircean standpoint, which means that it is not always related to rationality. All thinking is in signs, which can be icons, indices, or symbols, according to Peirce's philosophical and semiotic point of view. Additionally, all inference is a type of sign activity, where the word sign includes "feeling, image, conception, and other representation" (Peirce, 1866–1913, 5.283), or, to put it in Kantian terms, all synthetic forms of cognition. In this sense, the term "inference" refers to cognitive activity engaged in manipulative and model-based cognition as well as conscious processes. This concept of inference's broad meaning is also connected to my eco-cognitive model of abduction. In this model, cognition is understood in relation to an embodied subject who interacts with his surroundings, meaning that he receives and perceives information but also manipulates it, either directly or by using the creation of artificial entities. In this sense, the term "inference" does not only refer to conscious processes but also deals with cognitive activities involved in model-based and manipulative cognition (Magnani, 2009).

1 The nature of eco-cognitive situatedness determines the type of abduction at play

1.1 Data as suitable affordances that prompt abductive cognition: "ecological validity"

According to Gibson (1979), "affordance" is defined as what the surroundings furnish, offer, or produce. A chair, for example, provides the ability to sit, breathe in the air, swim in water, climb stairs, and more. The concept of agent-environment mutuality is referred to by affordances, which transcend the boundary between the subjective and objective. In addition to giving precise examples, Gibson also included a list of definitions (Wells, 2002) that can lead to possible misunderstandings:

1. affordances are chances for action;

2. affordances are the values and meanings of entities which can be directly perceived;

3. affordances are ecological events;

4. affordances point toward the mutuality of perceiver and environment.

The link between affordances and abduction (that is reasoning to hypotheses) is the subject of our concern in this subsection. Both human and non-human animals may "modify" or "create" affordances by adjusting their cognitive niches,[2] which can either facilitate or hinder particular abductive outcomes. Even the most fundamental and ingrained perceptual affordances accessible to our ancestors were likely considerably different from those we have now. It is also evident that human, biological bodies themselves develop and of course, children and all other animals exhibit a variety of affordances as well.

In his studies, Gibson essentially defined "direct" perception as the absence of an agent's internal inferential mediation or processing. In this sense, affordances—and the direct, uncomplicated way in which an organism takes them in—express the complementary nature of an organism and its environment (Wells, 2002). It is noteworthy to highlight that Gibsonian affordance as originally defined by Donald Norman is modified to include mental/internal processing: "I believe that affordances result from the mental

---

[2]The cognitive human acts that convert the natural world into a cognitive one are known as representational delegations to the external environment that are configured as elements of *cognitive niches* (some of which may be seen as pregnances; see Magnani, 2022, Lexicon of Discoverability). According to research conducted in the field of biosciences of evolution by Odling-Smee, Laland, and Feldman (Odling-Smee et al., 2003; Laland & Sterelny, 2006; Laland & Brown, 2006), humans have created enormous cognitive niches that are characterized by informational, cognitive, and ultimately computational processes.

interpretation of things, based on our past knowledge and experience applied to our perception of the things about us" (Norman, 1988, p. 14). It is possible for an event or location to offer distinct affordances to distinct organisms, while also providing many affordances to the same creature. According to Donald Norman, affordances indicate a variety of possibilities. Since artifacts are complicated entities and their affordances typically need extensive supporting data, it is more beneficial to examine them from this angle. For instance, understanding a door's complete range of affordances necessitates knowing intricate details like, say, the pull's specific direction of operation (Scarantino, 2003, pp. 953–954). Of course, among the many opportunities provided by affordances are some that are somewhat likely to provide a substantial foundation for human discovery, such as in the field of science.

As I have indicated previously, going beyond Gibsonian direct perception, higher representational and mental processes related to thinking and learning are frequently required in order to become attuned to invariants and disturbances present in the environment. For instance, when creating an artifact with the intention of accurately and beneficially displaying its entire range of affordances, we must distinguish between two levels: (1) the creation of the object's utility and (2) the defining of the potential (and accurate) perceptual cues that characterize the affordances that the artifact can offer. They are quite simple for the user/agent to complete (Gaver, 1991; Warren, 1995; McGrenere & Ho, 2000): "In general, when the apparent affordances of an artifact match its intended use, the artifact is easy to operate. When apparent affordances suggest different actions than those for which the object is designed, errors are common and signs are necessary" (Gaver, 1991, p. 80). In this last case affordances are apparent because they are simply "not seen". Information, as we know, frequently includes higher cognitive faculties and goes beyond what can be obtained by direct perception, arbitrating the perceivability of affordances in this way.

Like in manipulative abduction[3] and other less skilled and creative cases, where the resources are not just inner (neurally-specified) and embodied but also hybridly entwined with the environment, online thinking represents a true case of distributed cognition. In this case, we are dealing with an abductive/adaptive process produced in the dynamical inner/outer coupling

---

[3]To give a clear example, the idea of manipulative abduction captures a significant portion of scientific thinking in which the role of action and external models (such as for example diagrams and artifacts) and devices is central, and in which the characteristics of this action are implicit and difficult to elicit. It also considers the external dimension of abductive reasoning from an eco-cognitive perspective. Action can supply knowledge that would not otherwise be available, allowing the agent to initiate and carry out an appropriate abductive process of hypothesis development and/or selection. We have to further say that manipulative abduction occurs when we are "thinking through doing" and not only, in a pragmatic sense, about doing (Magnani, 2009, chapter one).

where internal elements are "directly *causally* locked onto the contributing external elements" (Wheeler, 2004, p. 705).

According to Brunswik's hypothesis, an organism must *infer* information about what is happening in its ecological niche from the cues that are accessible, which are supplied by proximal stimuli, rather than being able to directly sense distant stimuli. The ecological validity of this "vicarious" inference, according to Brunswik, is, of course, compromised by the very changeable diagnostic nature of the accessible signals as well as their inherent incompleteness, unreliability, ambiguity, and equivocality. Implicitly expressing an abductive attitude commensurate with Peirce, Brunswik says: "[...] ordinarily organisms must behave as if in a semierratic ecology" (Brunswik, 1955, p. 209), considering the inherent "ambiguity in the causal texture of the environment" (Brunswik, 1943, p. 255). He continues by saying that in this sense, both the cues and the mediated inference are "probabilistic," much as in an abduction scenario where it is always the case that: "Both the object-cue and the means-end relationship are relations between probable partial causes and probable causal effects" (Brunswik, 1943, p. 255).

Accordingly, the Brunswikian notion of *ecological validity* may be understood in terms of the inference's abductive plausibility in light of the relevant information and cues; in other words, ecological validity and the concepts of discoverablity and diagnosticability are congruent. The degree of adaptation between an organism's behavior and the environment's structure is measured by the quality of the inferential abductive performance or the fitness of the behavior based on the specific chosen inference. The scenario is similar to what I have described in the instance of the so-called "visual abduction" when the cues are the subject of an easy and quick perceptual evaluation (Magnani, 2009, chapter two).[4] In contrast, in the other scenarios, organisms more or less correctly inferentially make a "hypothesis/judgment" on the environment's distal structure. Again, this viewpoint makes Gibson's intuition easier to understand: "Perceiving is the simplest and best kind of knowing" (Vicente, 2003, p. 261).

However, there are further types. Using instruments to learn expands perception into the domain of the very small and the very far away; using language to learn makes knowledge explicit rather than implicit (Gibson, 1979, p. 263). An illustration of this would be a forecast of wind behavior, which is often probabilistic and reliant on the current wind speed recorded at a ground station and shown on a computer screen as the "cue." It is noteworthy to mention that in this particular instance, the day-after action of dressing appropriately for the weather is made possible by the proximate perception.

---

[4]In this last instance, we may state that the proximal and distal structures are mapped one to one (Vicente, 2003, p. 261).

Studies grounded in the Brunswikian tradition have highlighted the fundamentally ecological nature of the cognitive engineering endeavor within the context of systems made up of human interaction, humans, mediating technologies, and tasks environments. Numerous findings have demonstrated in a variety of fascinating ways how technology gadgets support humans in fulfilling their environmental adaptability by improving the creation of hypotheses, judgment, and, ultimately, decision-making. Sometimes the technology itself is unable to make the best decision about a particular scenario, and other times the interaction between the user and the technology introduces a gap in the proximal/distal connection (Kirlik, 2006b).

Understanding perception and other cognitive processes as methods of locating important information using extra-neural active processes associated with the body and social environment brings back the concepts of cognitive activity and its "situatedness", which I have recently discussed in my studies (Magnani, 2022). It is a way of getting more sensory data, compensating for their equivocality, and reaching cognitive feedback, and/or a way of manipulating them, and also of exploiting cognitive delegations to the environment and to artifacts. Thus, brains do not need to store information since they do not need to create intricate internal representations of their surroundings.

1.2 The plasticity of environmental situatedness. Affordances, diagnosticability, and creative abduction

As I said before, Gibson was certain that "The hypothesis that things have affordances, and that we perceive or learn to perceive them, is very promising, radical, but not yet elaborated" (Gibson, 1979, p. 403). To delve further into this matter, we may argue that the very fact that a chair allows one to sit implies that we are able to identify certain cues (stiffness, rigidity, and flatness) that make it simple for someone to state, "I can sit down." Assume that the same individual now possesses item O. Here, the individual is limited to perceiving its flatness. For example, he or she has no idea if it is sturdy and stiff. Nevertheless, he or she chooses to sit down on it and manages to do so. The issue of direct and indirect visual perception arises once more. We are able to identify and stabilize the new affordances because of the action impact.

My point is that we need to make a distinction between the two situations. In the first, the indicators we identify—flatness, robustness, and rigidity—are very diagnostic for determining whether or not we can sit down on it. In the second, on the other hand, we ultimately decide to sit down but lack specific information about it. How many flat objects are there that are not suitable for sitting on? Although a nail head appears flat, sitting on it is not recommended. This illustration makes two crucial points very clearer: first off, creating affordances is a (semiotic) inferential process (see Windsor,

2004); second, it emphasizes the connection that exists in the eco-cognitive interplay between an organism's environment and the knowledge that defines it. In the last instance, information is obtained by a straightforward action; in other instances, it requires an action and intricate manipulations.

"Highly diagnostic" relates specifically to the abductive framework. In the first chapter of my book on abduction (Magnani, 2009), I defined abduction as the process of *inferring* certain facts, rules, and hypotheses that make certain sentences tenable, and that *explain* or *discover* some (ultimately novel) phenomena or observation. From Peirce's philosophical perspective, I have said repeatedly that all thinking is in signs, which can be icons, indices, or symbols. Additionally, all inference is a type of sign activity, where the term sign encompasses "feeling, image, conception, and other representation" (Peirce, 1866—1913, 5.283), and, in Kantian words, all synthetic forms of cognition. In other words, a significant portion of the cognitive process is "model-based" and, as a result, non-sentential. Naturally, when model-based reasoning is integrated into abductive processes, it takes on a unique and creative significance that allows us to identify a *model-based abduction*. When doctors uses diagnostic reasoning, for example, if they see several symptoms (signs or clues) in several ways, such as fever, chest discomfort, and cough, they may conclude that the patient has pneumonia.

As I already said, the original Gibsonian concept of affordance focuses mostly on situations where the "perceptual" cues and indicators that we are able to recognize prompt or indicate one course of action over another. They already exist and are typical examples of how an organism adapts to a particular ecological niche. On the other hand, affordances may be linked to the variable (degree of) abducibility of a configuration of signs if we accept that environments and organisms have to exploit both instinctual and cognitive plastic endowments. For example, a chair facilitates sitting in the sense that sitting is a sign activity in which we perceive certain physical properties (flatness, rigidity, etc.), and as such, we can typically "infer" (abduce, in the Peircean sense) that a possible way to cope with a chair is sitting on it. Put another way, because affordances are pre-existing in the perceptual and cognitive endowments of both human and non-human animals, it is, for the most part, a spontaneous abduction to locate them.

In my opinion, explaining affordances in this way could help to make sense of some of Gibson's puzzling themes, particularly the assertion that humans directly perceive affordances and that object's value and meaning are immediately apparent. Organisms possess a standard set of affordances (such as those derived from their hardwired sensory system),[5] but they can

---

[5]The word "wired" is prone to misunderstandings. I generally agree that there are two types of cognitive aspects: "hardwired" and "pre-wired". I mean by the former word the parts of cognition that are predetermined and cannot be changed. On the other

also expand and alter the range of what is available to them by using the appropriate cognitive abductive skills. I also emphasize how crucial it is to remember that as environments change, so do the perceptual capacities enhanced by new or higher-level cognitive skills—that is, those capacities beyond those granted by merely instinctual levels. Although affordances are typically stabilized, this does not mean that they cannot be altered or replaced, nor that new ones cannot be formed.

Because affordance perception is abductive, it primarily depends on a cognitively-related, ongoing process of hypothesis-making. That A affords B to C can be also considered from a semiotic perspective as follows: A signifies B to C. A is a sign, B the object signified, and C the interpretant. Cognitive skills related to a particular domain (such as knowledge contents and inferential capacities, but also appropriate pre-wired sensory endowments) allow the interpretant to make certain abductive inferences from signs (such as perceiving affordances) that are not possible for those without those apparatuses. To ordinary people a cough and chest pain are not diagnostic, because they do not know what the symptoms of pneumonia or other diseases related to cough and chest pain are. Thus, they cannot make any abductive inference of this kind and so perform subsequent appropriate medical actions.

2 Discoverability and diagnosticability through affordance creation

Think of a large stone and a chair, for example. Both of these items do, in fact, allow for sitting. The distinction lies in the fact that affordances in the instance of a stone are essentially presumptive: we typically "infer" (in the Peircean sense) that a stone may be beneficial for sitting when we come across one. On the other hand, chairs' are *manufactured* in some way from scratch. In the instance of a chair, we have fully created an entity that exhibits a range of affordances. Using the abductive paradigm we presented above, this affordance creation process may be better understood.

When an entity allows us to do a specific action, it implies that it incorporates the signs that allow us to "infer"—through a variety of acquired and instinctive cognitive endowments—that we may engage with it in a particular way. As mentioned previously, when it comes to stones, humans take advantage of an already-established configuration or structure of meaningful sign data that has been shaped by past evolutionary experiences with the human body (and, to some extent, by "material cultural" evolution, such as

hand, the latter term describes those skills that are built prior to the experience, but that are modifiable in later individual development and through the process of attunement to relevant environmental cues. This distinction helps to highlight the significance of development and how it relates to plasticity. Genes and inbuilt components do not predetermine every facet of cognition. See further Barrett & Kurzban (2006).

when hominids used a stone or chair to sit in front of a primitive altar). In the case of a chair, this configuration is invented. If this viewpoint is valid, we may contend that creating "artificial" affordances entails configuring signs in the outside world specifically to create new, accurate inferences of affordability. By doing this, we carry out deft manipulations and acts that, I conjecture, might provide new (and sometimes "unexpected") affordances. Therefore, affordance creation also entails making new ways of inferring them feasible: a process that is fundamentally tied to improving discoverability and diagnosticability.[6]

2.1 Manipulating external representations and artifacts to create chances

It is now evident that the development of culture, artifacts, and technologies across time may be seen as an ongoing process of creating new affordances on top of or even starting from scratch. Humans and the environments they have created and inhabited have coevolved from cave art to contemporary computing. In fact, when compared to the prospects and chances offered by other tools and technology, what a computer may afford encompasses an astounding diversity of possibilities. More specifically, a computer may mimetically duplicate a significant portion of the most sophisticated operations that the human brain-mind systems can do, acting as a Practical Universal Turing Machine (see Turing, 1992 and Magnani, 2021) (Magnani, 2006). For example, computers are even more powerful than humans in several ways, such as memory and certain areas of mathematical thinking. From a semiotic standpoint, computers bring into existence new artifacts that offer and create new affordances—that is, they present "signs" in the Peircean sense for exploring, expanding, and manipulating our own brain cognitive processing. In this way, they help to "extend the mind beyond the brain." Building affordances, as was previously said, is primarily an abductive semiotic activity in which cues are placed strategically across the environment to promote a certain interaction above others.

To understand this better, think of basic diagrammatic demonstrations of rudimentary geometry—something we have all learned to perform in middle school—as the archetypal example of manipulative abduction. Additionally, they are instances of how affordances from the field of elementary geometry can be constructed so that, in the case of current learners, they can aid

---

[6]I have demonstrated an analogous problem in (Magnani, 2007): a lot of objects operate as "moral mediators." This phenomenon occurs when manipulations of artifacts and interactions among agents at a local level, such as in the case of the internet's effect on privacy or the derivatives in the global economic crisis, lead to macroscopic and increasingly prevalent global moral consequences on collective responsibilities. For instance, individuals' manipulations on the internet may have unnoticed effects on other people's privacy.

in reaching the conclusion of a proof, and, in the case of ancient pre-Euclidean geometers, they provided the necessary discoverability to yield new geometrical results.

In these situations, new visual affordances are revealed through the so-called diagrammatic *constructions*, which result from the straightforward modification and complication of appropriately externally shown diagrams. In order to readily arrive at generalized results—which, in the case of an axiomatic organization of elementary geometry, are termed theorems – the process involves building and modifying initial suitably depicted external diagrams. If the process is viewed as a broad inference leading to a result through a problem-solving exercise, it involves a distributed interaction between a continuous externalization through cognitive acts, its manipulation, and re-internalization that recognizes what has been learned from the outside, picking up the result and reinternalizing it. New affordances in the action materialize and lead to the outcome.

From an epistemological perspective, the situation shown above is a classic case of the manipulative abduction I mentioned before. Reframed in terms of affordances, this is a cognitive manipulation (completely abductive, the goal is to find an incontrovertible geometrical hypothesis – new or already known) in which an agent organizes epistemic actions that structure the environment in a way that unearths new affordances as opportunities that favor new outcomes when confronted with merely "internal" representational geometrical "thoughts," from which alone it is difficult or impossible to extract new meaningful features of a mathematical concept. As previously stated, affordance detection is hypothesis-driven; it is not claimed that everyone can do so. Only someone who has studied geometry can deduce the affordances within the manipulated construction built upon the original diagram. Thus, affordances that are deemed "artificial" are closely linked to the culture and knowledge that are accessible inside particular cognitive niches of humans, as well as to the appropriate individuals engaged in the process of epistemic inquiry.

The construction of a diagram offers nested affordances:

1. it is a straightforward image that virtually everyone, including many animals, can perceive and comprehend as a perceptual image that offers potential colors and shapes based on the perceptual hardwired endowments of the organism in front of it, such as cats and uneducated people (strict Gibsonian case);

2. it is an image that, with all of its technical characteristics, can be viewed and comprehended as a geometrical diagram (in this instance, a higher level of cognitive ability is required in the creature in question);

3. it is an artifact that can provide new affordances to be absorbed and perhaps added to the existing library of geometrical knowledge through even more inventive cognitive manipulations. Consider a young student who is required to "demonstrate" a simple geometry theorem, such as the sum of a triangle's interior angles. Since this theorem has previously been found (demonstrated) historically and is documented in all Euclidean geometry manuals, the youngster does not need to prove it for the first time. With the exception of the scenario in which he repeats the proof by rote, he may accomplish this demonstration, however, by employing the sequence of suitably extracted affordances, which are predicated on the sensible application of fundamental geometric ideas that he is already familiar with. We may also argue that the youngster employed a heuristic, which is a sophisticated method of investigation. Naturally, this heuristic is a real "demonstration" and plainly does not result in discovery when seen through the lens of an existing geometry handbook (as an abstract and static system of knowledge). It is, instead, a sort of "rediscovery". It is a re-discovery from the perspective of the child-subject as well, as he finds a property that was first granted to him. Rather, the inferences made at the time of the initial historical discovery (perhaps Greek) of that triangular attribute and the evaluation of the corresponding theorem produced a sort of creative achievement (a creative manipulative abduction, as I have stated). Furthermore, as both kinds of reasoning rely on "hybrid" forms of representation that include significant non-verbal cues (like geometric diagrams), they are primarily model-based as well.[7]

Because animals, infants, and adults all have different perceptual endowments, they can all perceive "the brink of a cliff as *fall-off-able* according to a common perceptual process" (Scarantino, 2003, p. 960) which explains why affordances can be grasped simultaneously by all three cognitive differences: "This is much the same as we would describe a piano as having an affordance of music playability. Nested within this affordance, the piano keys have the affordance of depressability" (McGrenere & Ho, 2000, p. 340). It is also possible to add that the piano provides chance—discoverability—in the cognitive interplay artifact/agent, offering fresh affordances of new good melodies, not previously generated in a merely internal/mental way, in the musician's mind, but found over there, in the hybrid interplay with the musical artifact. Of course, depending on their qualities, degrees of affordance, and other characteristics, the diagram and the piano, as well as other

---

[7] Of course, the agent can alter the artifacts' characteristics in a more or less inventive manner in order to increase the visibility or exploitability of the affordances that are already there or to create new ones that are provided as choices. An instance of this would be if a user alters a computational interface by creating an alias for an extended command string. Instead of writing a lengthy string of characters, he or she can utilize the tool more easily by hitting a single key or many keys at once (McGrenere & Ho, 2000).

artifacts, present different limited conditions for affordances. The example above can be explained in terms of variables and proximal/distal distinctions according to Kirlik's perspective. The agent creates a diagram in which he or she can operate on the surface by utilizing the constraints that ensure that latent variables inherent to the materiality of the artifact at hand "take care of themselves, so to speak" (Kirlik, 2006a, p. 221).

Because different aspects are released from the agent and assigned to the external representation, which provides a proximal perceptual and manipulative environment with all the resources required to successfully carry out the creative task of finding new answers to a certain geometrical question, the need for having a rich internal model of a depicted geometrical diagram is weakened. Since the outcome is readily apparent, it may be taken up and internally reinterpreted. The manipulation of the figure, which is a model in the dynamics of geometrical thinking, demonstrates a situation in which cognition and perception are fully integrated.

From a semiotic point of view, we do not initially possess the cognitive capacities necessary to internally infer the solution of the problem. By modifying the externalized configuration, or the external diagram, we are able to create a new perceptual sign configuration with attributes that were not present in either the original external or internal representation. We are able to solve the problem thanks to a new set of affordances that are created by this altered sign arrangement. As we have said, it is also a means of "demonstrating" a new theorem in the Euclidean sense. This example provides an epistemological illustration of the nature of the cognitive interplay between external representations and internal neuronal semiotic configurations that enable representational thought about an initial problem (along with the aid of various embodied "cognitive" kinesthetic and motor abilities): additionally, also for Peirce, more than a century before the new ideas provided by the studies on distributed reasoning, the two aspects are pragmatically intertwined.

The "hypothetical" nature of affordances serves as a reminder that it is not necessarily the case that just anybody can detect it, affordances are only potentialities for organisms. First of all, the ability to perceive affordances stems from an abductive process in which we infer potential strategies for interacting with an entity based on the signs and clues that are at our disposal. I have to repeat, a portion of affordances is relatively constant, pre-specified or neurally encoded in the perceptual system. These affordances are referred to as "invariants," using a word from physics that Gibson also uses to describe affordances with a strong cognitive valence Because of our cognitive-biological configuration, which makes it easy for humans to acquire the corresponding cognitive ability as a "current" and "reliable" ability, perceiving the affordances of a chair is in fact rooted and "stabilized" in our cultural evolution (Scarantino, 2003, p. 959).

We stated that the majority of the distinctions that we can recognize are, in a sense, *intra-species*. Since intra-species variations appear to be heavily implicated, there is something unusual in the high-level cognitive performance on a geometrical concept and its figure. For example, only someone who has studied geometry can infer (and so "perceive") the affordances "inside" the newly constructed, altered structure that is based on an original geometrical problem. This relates to the "expertise" issue I mentioned earlier. Firstly, there is a close relationship between manufactured affordances and culture and social contexts. Second, affordances have to do with education. Certain affordances, like those of a geometrical construction, may be taught and acquired once they are created; in fact, perceiving them is not an innate ability. Recognizing this fact, of course, emphasizes even more the dynamic character of affordances in organisms' plastic cognitive life, beyond their evolutionary character.

In sum, the ability to execute clever manipulations is related to the process of generating external representations. According to Donald (2001) and myself (2009, chapter three), humans are always involved in cognitive mimetic and creative processes that include representing their ideas, solutions, and thoughts into appropriate external structures and products. By doing this, individuals produce outward representations of certain internal, adequately stored propositional and model-based representations that are already available within their brains. Sometimes, these external counterparts can be creatively employed to find space for new concepts and new ways of inferring that cannot be exhibited by the mere "internal" representation alone. Initially, these external counterparts merely mirror ideas or thoughts already present in the mind (Magnani, 2006). When humans construct these external representations (which, I repeat, might be viewed as essentially mimetic but can also become "creative"), they alter the environment in a way that uncovers new cognitive opportunities and improves *discoverability* and *diagnosticability*.

By doing this, new affordances are created and made collectively available. In a broader sense, we might restate from this angle as well: abduction also involves the ongoing process of modifying the surroundings to provide affordances, or fresh opportunities for action.[8]

---

[8]Analysis of the so-called "adaptive problem solving" has been offered by an intriguing research on the function of models, which came from a 5-year empirical ethnographic study of two systems biology laboratories and their partnerships with experimental biologists (MacLeod & Nersessian, 2016).

3 Conclusion

I have argued in this chapter that discoverability naturally depends on the availability of suitable affordances. Individuals constantly distribute and assign cognitive functions to their surroundings in an effort to reduce their limits. They provide representations, models, and other kinds of mediating structures that are supposed to support discoverability and act as cognitive help. Regarding the utilization of cognitive resources integrated into the surroundings, I have emphasized the importance of affordances and the goal of enhancing the recently developed framework known as EEEE-cognition (extended, embodied, embedded, enacted). From this angle, I have gone on to explain human cognition and its evolutionary aspect in terms of environmental situatedness, where bodies and external, artifactual entities and gadgets play significant roles. Accordingly, fresh opportunities for discovery—to use my own terminology—become not only information but also "affordances," or contextual anchors that enable us to make the most use of outside cognitive resources. Insofar as it exemplifies those hypothetical conclusions that are driven by activities consisting of astute manipulations of the surroundings to both identify novel affordances and generate artifactual external objects that provide "novel" affordances/cues, abduction has remained in operation.

Acknowledgement. Parts of this chapter are excerpted from my monograph *Discoverability. The Urgent Need of and Ecology of Human Creativity*, Springer, Cham, 2022, chapter two. For the informative critiques and interesting exchanges that assisted me in enriching my analysis of the issues treated in this article, I am obligated to my collaborators Emanuele Bardone, Tommaso Bertolotti, Selene Arfini, and Alger Sans Pinillos.

References

Barrett, H. C., & Kurzban, R. (2006). Modularity in cognition: Framing the debate. *Psychological Review, 113(3)*, 628–647.

Brunswik, E. (1943). Organismic achievement and envornmental probability. *Psychological Review, 50*, 255–272.

Brunswik, E. (1955). Representative design and probabilistic theory in a functional psychology. Psychological Review, 62, 193–217.

Donald, M. (2001). *A mind so rare. The evolution of human consciousness*. London: Norton.

Gaver, W. W. (1991). Technology affordances. In *CHI'91 Conference Proceedings* (pp. 79–84).

Gibson, J. J. (1979). *The ecological approach to visual perception*. Boston, MA: Houghton Mifflin.

Kirlik, A. (2006a). Abstracting situated action: Implications for cognitive modeling and interface design. In A. Kirlik (Ed.), *Human-technology interaction. Methods and models for cognitive engineering and human-computer interaction* (pp. 212–226). Oxford/New York: Oxford University Press.

Kirlik, A. (Ed.). (2006b). *Human-technology interaction. Methods and models for cognitive engineering and human-computer interaction*. Oxford/New York: Oxford University Press.

Laland, K. N., & Brown, G. R. (2006). Niche construction, human behavior, and the adaptive-lag hypothesis. *Evolutionary Anthropology, 15*, 95–104.

Laland, K. N., & Sterelny, K. (2006). Perspective: Seven reasons (not) to neglect niche construc- tion. *Evolution. International Journal of Organic Evolution, 60(9)*, 4757–4779.

MacLeod, M., & Nersessian, N. J. (2016). Interdisciplinary problem-solving: Emerging modes in integrative systems biology. *European Journal for Philosophy of Science, 6(3)*, 401–418.

Magnani, L. (2006). Mimetic minds. Meaning formation through epistemic mediators and external representations. In A. Loula, R. Gudwin, & J. Queiroz (Eds.), *Artificial cognition systems* (pp. 327–357). Hershey, PA: Idea Group Publishers.

Magnani, L. (2007). *Morality in a technological world. Knowledge as duty*. Cambridge: Cambridge University Press.

Magnani, L. (2009). *Abductive cognition. The epistemological and eco-cognitive dimensions of hypothetical reasoning*. Heidelberg/Berlin: Springer.

Magnani, L. (2021). Computational domestication of ignorant entities. Unconventional cognitive embodiments. *Synthese, 198*, 7503–7532. Special Issue on "Knowing the Unknown" (guest editors L. Magnani and S. Arfini).

Magnani, L. (2022). *Discoverability. The urgent need of and ecology of human creativity*. Springer, Cham.

McGrenere, J., & Ho, W. (2000). Affordances: Clarifying and evolving a concept. In *Proceedings of Graphics Interface* (pp. 179–186). May 15–17, 2000, Montreal, Quebec, Canada.

Norman, D. A. (1988). *The psychology of everyday things*. New York: Basic Books.

Odling-Smee, F. J., Laland, K. N., & Feldman, M. W. (2003). *Niche construction. The neglected process in evolution*. Princeton, NJ: Princeton University Press.

Peirce, C. S. (1866–1913). *Collected papers of Charles Sanders Peirce*. Vols. 1–6, C. Hartshorne & P. Weiss (Eds.), vols. 7–8, A. W. Burks (Ed.). (1931–1958). Cambridge, MA: Harvard University Press.

Scarantino, A. (2003). Affordances explained. *Philosophy of Science, 70*, 949–961.

Turing, A. M. (1992). In D. C. Ince (Ed.), *Collected works of Alan Turing. Mechanical intelligence*. Amsterdam: Elsevier.

Vicente, K. J. (2003). Beyond the lens model and direct perception: Toward a broader ecological psychology. *Ecological Psychology, 15(3)*, 241–267.

Warren, W. H. (1995). Constructing an econiche. In J. Flach, P. Hancock, J. Caird, & K. J. Vicente (Eds.), *Global perspective on the ecology of human-machine systems* (pp. 210–237). Hillsdale, NJ: Lawrence Erlbaum Associates.

Wells, A. J. (2002). Gibson's affordances and Turing's theory of computation. *Ecological Psychology, 14(3)*, 141–180.

Wheeler, M. (2004). Is language an ultimate artifact? *Language Sciences, 26*, 693–715.

Windsor, W. L. (2004). An ecological approach to semiotics. *Journal for the Theory of Social Behavior, 34(2)*, 179–198.

On scientific creativity and its limitations

Fabio Minazzi

Dipartimento di Scienze Teoriche e Applicate, Università degli Studi dell'Insubria, Via O. Rossi 9, Padiglione Rossi, 21100 Varese, Italy

> Between the idea and the realization of the intention always lies a period of work and effort typical of the inventive process. [...] The arising of the idea is that happy moment in the creative activity of thought in which everything seems possible, since it has nothing to do with reality yet. Execution is the moment when one has to procure all the means necessary for the realization of the idea, a moment still creative, still happy, a moment when one has to overcome the resistances of nature; from it one always emerges tempered and ennobled, even if beaten.
>
> Rudolf Diesel, *Die Entstehung des Dieselmotors*, Berlin 1913, pp. 151–152

1 Modern science was born in the seventeenth century

Science was born in the seventeenth century, the age of the Baroque. Is this an accident? I do not think so. In many historical reconstructions, especially those devoted to Italian history, there is an insistence that the seventeenth century constituted an age of crisis. If it was a time of "crisis", however, it was a very strange "crisis". For what reason? Because with the birth of modern science a genuine point of no return was introduced into the history of all mankind. Not for nothing did an English historian such as Herbert Butterfield in his *The Origins of Modern Science* (1958) argue that the birth of science constituted a genuine turning point:

> Since that revolution overturned the authority in science not only of the middle ages but of the ancient world—since it ended not only in the eclipse of scholastic philosophy but also in the demolition of Aristotelian physics—it outshines everything since the rise of Christianity and reduces the Renaissance and Reformation to the level of mere episodes, mere internal displacements within the system of medieval Christendom. Since it changed the character of men's habitual mental operations even in the conduct of the non-material sciences, while transforming the whole diagram of the physical universe and the very texture of human life itself, it looms so large as the true source of the modern world and of the modern mentality that our customary periodisation of European history has become an anachronism and an encumbrance.[1]

This approach was largely shared by an Italian epistemologist such as Ludovico Geymonat, who in his monumental *History of Philosophical and*

---

[1] H. Butterfield, *The Origins of Modern Science*, G. Bell & Sons Ltd, London 1958, pp. VII–VIII.

Scientific Thought (in 7 vols.) precisely assumed the birth of modern science as the authentic and fundamental turning point in human history. So much so that by leveraging this strategic "Archimedean" point, Geymonat overturned the traditional historiographical framework of the main *Histories of Thought*, devoting 5 volumes of his great *History* to the study of modernity and the contemporary era. Geymonat also always considered scientific thought, entwined with philosophical thought, as the privileged *fil rouge* of his historical reconstruction. Thus, as his *History* approaches the present era, the volumes expand and deepen and are dilated, to the point where two very large volumes are devoted to twentieth-century thought, flanked by another tome for the study of the transition between the nineteenth and twentieth centuries. In contrast, only one volume, the first, is devoted to ancient and medieval thought. This critical reversal of the traditional historiographical approach (which, in general, pays increasing attention to the centuries of the distant past, gradually reducing its focus on the contemporary age) rests on the conviction that, since science first developed in the seventeenth century, nothing has been the same as before. In the words of Bertrand Russell in *The Scientific Outlook* (1931), "one hundred and fifty years of science have proved more explosive than five thousand years of prescientific culture".[2]

In this reconstructive context, the acknowledged father of modern science is, without a doubt, Galileo Galilei. He provided the first outline of modern science, working at the turn of the 16th and 17th centuries, although it was in the latter century that his most important and significant works appeared. Notably, his scientific masterpiece was published just when Galileo had already been condemned, following the famous inquisitorial trial to which he was subjected, since his *Discourses and Mathematical Demonstrations Relating to Two New Sciences* was published anonymously in Leyden by Elzevier in 1638. If Galileo had not published this scientific treatise, he would not be considered the acknowledged father of modern science. If anything, he would have been remembered as a scholar who contributed, in a very significant and timely way to defending the Copernican theory, favouring its affirmation. In this key, Galileo would thus resemble Giordano Bruno as a courageous advocate of the new heliocentric and heliostatic astronomical theories. On the contrary, the very publication of the *Discourses* qualifies Galileo as the first scientist of modernity who made a decisive contribution to the birth and development of modern science. In fact, the two sciences discussed in the *Discourses* are, precisely, the resistance of materials and dynamics, the science of motion, i.e., that of a rigid body moving at a speed significantly less than the speed of light.

---

[2]Bertrand Russell, *The Scientific Outlook*, George Allen & Unwin, London 1931, p. 9–10.

2 Baroque science and culture: imagination and sense of reality

According to Fernand Braudel,[3] Baroque culture constituted a new form of taste, culture and even 'civilisation' that spread from Italy to the whole of Europe, helping to create modern theatre, opera and modern science. But how can the specific contribution of Baroque culture be specified? I would say that Baroque culture insisted on two different and yet interrelated moments, namely the sense of reality and the role of imagination. Thus, on the one hand, there is a kind of profound reaction to the traditional balance of classical rationality, which was now intertwined with an eccentric, bizarre, paradox-loving creativity, but, on the other hand, precisely within this same unbridled imagination, the almost infinite complication of reality is also recovered. Precisely because—in the words of an eminent writer like Carlo Emilio Gadda—the world is very complex and always entangled, precisely because it resembles a ... dumpling.

Well, precisely this unique intertwining of imagination and sense of reality is also found within the Galilean image of science. This statement may seem provocative, since we are often conditioned—willingly or unwillingly—by an empiricist image of science, according to which scientific knowledge is, in the final analysis, a derivative and a product of sense and experimental experience itself. This traditional image of science has also largely dominated epistemological thought, so much so that the tradition of empiricism—from Hume's to the Viennese brand of logical empiricism—has provided a privileged frame of reference for trying to construct a correct image of scientific knowledge. However, it is precisely the historical and conceptual hegemony of this albeit fruitful epistemological research program, based on empiricism, that has contributed to the diversion from a correct image of scientific knowledge. The latter does not derive so much from experimental experience by inductive means, for it arises, if anything, from a complex interplay that is more highly articulated and also much more fruitful.

If one considers the various reactions aroused in Galileo's contemporaries by the reading of his *Discourses*, it is easy to understand this problem concerning both the genesis of modern science and its epistemological structure. When Aristotelian physicists read Galileo's *Discourses* and his innovative treatment of the motion of rigid bodies, there was no shortage of direct reactions. Such as that of the Genoese physicist Giovanni Battista Baliani, the author of a book *De motu naturali gravium solidorum* (1638) in which this physicist had reached, by experimental means, the correct determination of the law that regulated falling bodies. For this reason, Baliani in his correspondence with Galileo insisted on emphasizing the decisive role of

---

[3]See F. Braudel, *Out of Italy: Two Centuries of World Domination and Demise*, Europa Editions, London 1994, pp. 66–67.

experience, which, in his opinion, also derived from almost all of Galileo's work, in deep harmony also with Aristotle's teaching, aimed at putting experience before our theories, as Galileo himself repeatedly emphasized in his *Dialogue Concerning the Two Chief World Systems* (1632). In fact, Baliani (in a letter dated 1 July 1939) wrote: "I in truth have judged that experiences should be taken as principles of science, when they are certain, and that from things known by sense it is part of science to lead us into cognition of the unknown" (XVIII, 69).[4]

Before this privileged appeal to the foundational role of experience, Galileo, in his earlier letter of 7 January 1639, had, however, already written to Baliani (thanking him for sending him his book on motion) noting that he had

> himself dealt with that subject [of motion—*ed.*], but in a much more extensive way and with a different approach. This is due to the fact that I do not admit as a hypothesis anything except the definition of motion, which I wish to deal with by demonstrating its accidents, in this imitating Archimedes on Spirals. (XVIII, 11–12).

In other words, Galileo, while claiming that he also dealt with the problem of the motion of bodies, stresses that he nevertheless followed a different path. In fact, Galileo, unlike Baliani, did not start from the study of experience, but rather from some definitions that he introduced *ex suppositione*, that is as a hypothesis, merely conjecturally. In carrying out this "different attack" of his, Galileo declares, however, that he is referring back to the great scientific model of Archimedes of Syracuse, who, in *De Spiralibus*, also started from some hypothetical definitions, caring little whether or not these geometric forms existed in the world. For this reason Galileo also specifies the following:

> but returning to my treatise on motion, I argue *ex suppositione* regarding motion, defined in that way; so that even if the effects did not correspond to the accidents of the motion of descending weights (bodies), it would matter little to me, just as nothing is lacking in Archimedes' demonstrations due to the fact that no moving body (moveable) is found to move in spiral lines. But in this I was, so to speak, fortunate, because the motion of bodies and their accidents correspond exactly to the accidents demonstrated by me relating to motion as I have defined it. (XVIII, 13).

In this way Galileo explains how, in order to construct his innovative theory of motion, he essentially proceeded in this way: first, he introduced

---

[4]This and all the other quotations from Galileo's writings that follow in the text are taken from the Edizione Nazionale of *Le opere di Galileo Galilei*, edited and directed by Antonio Favaro, G. Barbèra Editore, Florence 1968, 20 vols., The volume is shown in Roman numbers and the reference page in Arabic numbers. The translations are mine.

hypothetical, arbitrary definitions, which we might describe as absolutely conventional, precisely *ex suppositione*. Taking these conjectural definitions as a starting point, a theory is then constructed—by rigorously *deductive* means—which implies certain *consequences* and specific *predictions*. Finally, as the third constituent moment in this way of proceeding, it is precisely these consequences—that is, the "predictive" component of the theory— that are placed in close comparison with the *experimental dimension* in order to be able to confirm, or falsify, these same predictions. In this way, the articulated Galilean picture of scientific knowledge is not empiricist at all, but deductive and normative, because at its core it contains different moments that are interwoven with each other in an absolutely innovative way. With the consequence that the *physical object* is not then derived directly from experience, but is instead constructed and elaborated, in the first instance, by the imagination of the scientist, who on this very basis subsequently builds, by *deductive* means, a theory, the predictive *consequences* of which are finally checked and rigorously verified within the experimental dimension. Thus in the Galilean model of science there subsist different and opposing elements: that of the creative and conventional imagination, that of rigorous logical-mathematical deduction and, last but not least, the moment of verification—or experimental falsification. It is all of these different "moments" that, taken together, finally configure a scientific theory worthy of the name because it is capable—in the words of Leonardo da Vinci—of enabling us to grasp "a thread of truth". Therefore, even in the Galilean heuristic model, the experimental dimension undoubtedly plays a fundamental and decisive role, but it does not constitute the primary horizon from which one starts to construct a theory, because it has the equally fundamental and crucial role of being able to subject the theory that has been devised to a critical-experimental check. The experimental dimension thus plays a part that is not located at the beginning of the process (as Baliani argued in his treatise and letter of July 1639), because it plays, instead, its part precisely in the concluding and decisive phase of scientific reflection, that is, the one devoted to the rigorous experimental control of the different theoretical predictions.

The epistemological model thus outlined by Galileo is a decidedly counterfactual model. This confirms that for Galileo the image of knowledge is constructed by interweaving the imaginative capacity of the scientist with his vivid sense of experimentally investigated reality in order to be able to place theoretical predictions under critical scrutiny. This is why Galileo proudly claims to have followed an "aggressione diversa" (a "different approach"), wishing to clearly distinguish himself from Baliani's empiricism, which instead emphasises precisely the constructive role of valuing experience as the authentic foundation of knowledge. Within the plane of the history of epistemological thought, it is undoubtedly the case that it was precisely Baliani's inductivist and empiricist image that was very widely successful, while the Galilean one was gradually forgotten and removed. Which constitutes a curious fate, which can, however, be easily explained if one keeps in mind that empiricism has been, over the centuries, a highly privileged point of reference, especially for English-speaking culture and its various admirers. In this sense, the empiricist image of science has thus dominated unchallenged from the seventeenth century until the contemporary age. Of course, this is by no means to deny a definite role and function of the experimental dimension within the construction of scientific knowledge. But, precisely in light of the timely Galilean considerations, it is, if anything, a matter of knowing how to construct a more correct and articulated and truthful image of the specific ways in which scientific knowledge grows and is constructed in the praxis of scientific research.

3 Einstein and his conception of science

In the early 1950s, Albert Einstein was urged by his friend Maurice Solovine to illustrate his overall conception of scientific knowledge. In response to this question from a lifelong friend, Einstein, in his letter of 7 May 1952, submitted the following drawing:[5]

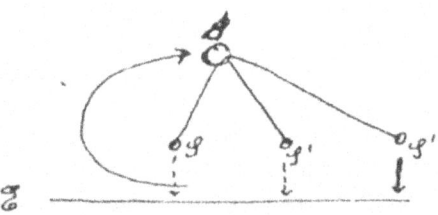

In this drawing, Einstein explains, the E line indicates the set of immediate experiences. That is, what we might refer to, with Husserl, as the *Lebenswelt*, the world of life, in which we all live as ordinary men and women because in this pragmatic dimension we are all present and active through the praxis we enact (and also undergo). This horizon common to all of us is, indeed, precisely the "world of life", which everyone—the scientist, the common man in the street, as well as the Nobel laureate—shares and within which they live and exercise their action as persons "of flesh and blood".

Point A, on the other hand, indicates axioms from which conclusions can be deductively drawn. In Einstein's drawing, the line leading to A is

---

[5]See A. Einstein, *Lettres a Maurice Solovine*, Gauthier-Villars, Paris 1956 and the English-language edition, A. Einstein, *Letters to Solovine*, The Philosophical Library, London 1994, where the letter quoted is on pp. 124–125.

represented by the arc of a parabola that appears slightly detached from line E. This "detachment" is precisely meant to emphasise the relative independence of the human imagination, which, while floating above the world of sensible perceptions, nevertheless manages to develop its own independent and autonomous path by which it comes to postulate axiom A. There is therefore, notes Einstein, no logical path leading from E to A. There is at most an intuitive connection that nevertheless enjoys its own relative autonomy within which the human imagination exercises its function.

It also emerges from the drawing that from the A's one can derive, this time by a rigorously deductive procedure, some particular utterances S that constitute the consequences of the theory and aspire to be true. But the verification—or falsification—of these utterances S occurs solely through comparison mediated by technology, which allows for the establishment of a decidedly experimental framework. However, Einstein points out, "this relation between the S's and the E's is nevertheless (pragmatically) much less uncertain than that which exists between the A's and E's (e.g., between the concept of dog and the corresponding immediate expressions). If such a correspondence, while remaining inaccessible to logic, could not be established with a high degree of certainty, the whole logical apparatus would be of no value for the purpose of 'understanding reality', e.g., theology)."

On the other hand, in the drawing, what joins the S's to the line E is indicated with a hatching that is intended precisely to emphasize the problematic character of this link that is established between theories and reality. Einstein writes: "The essential aspect here is the eternally problematic link between the world of ideas and what can be experienced (sensible experience)." Exactly in this problematic nexus between the theoretical plane of ideas and thoughts and its direct connection with the actual, real world lies that plane of scientific activity within which Galileo said that the scientist must be able to "climb over the impediments of matter". For this reason, as Rudolf Diesel also acknowledged in the quotation placed as an epigraph at the opening of this paper, "every inventor must be an optimist: the power of the idea retains all its active force only in the soul of its author, and only this one possesses the sacred fire of its realisation."[6]

In any case, all these elements—precisely represented by the E line, the A's, the S statements and the problematic link with the world of sense praxis mediated by the technological-experimental connection—configure the Einsteinian image of science. It is not difficult, however, to discern the profound similarity and close congruence that exists between this Einsteinian image of science and Galileo's illustrated in the previous section. Both of these two eminent physicists thus developed a complex and articulated image of scientific knowledge within which various elements—even decidedly

---

[6]Rudolf Diesel, *Die Entstehung des Dieselmotors*, Berlin 1913, p. 152.

contrasting—nevertheless play their own precise heuristic part to enable us, finally, to elaborate an objective knowledge of reality. Naturally this objective knowledge of reality can only be intrinsically problematic because, as mentioned earlier, it merely allows us to grasp a "thread of truth." In other words, this epistemological conception of scientific knowledge is never an absolute achievement, but always relative and critically appropriable. On the other hand, this outcome does not open up to any relativism precisely because we are in the presence of objective knowledge, which is always constructed within a well-defined and known context. For this reason, this objective knowledge, circumscribed to a finite and determinate physical sphere, is, within the limits of this finite configuration, also an "absolute" knowledge, that is, a knowledge that, within those determinate limits, allows us precisely to distinguish knowledge from lack of knowledge. We are thus in the presence of a historical-objective knowledge that by its nature always constitutes a historical-evolutionary knowledge that can be constantly critically extended in order to identify a deeper level of knowledge. On the other hand, this historical-evolutionary epistemological model prevents us from speaking of an absolute truth capable of providing us with exhaustive knowledge of the real. On the contrary, the latter is always critically capable of being deepened, because the game of human knowledge develops precisely through the ability to constantly question the results achieved in order to extend, step by step, our own knowledge of this "strange world" into which we have been catapulted from birth.[7]

---

[7] For a fuller critical study from this epistemological perspective, the reader is referred to my recent volume *Historical Epistemology and European Philosophy of Science*, Springer, Cham 2022.

Hypothetical value judgements: Reconciling value-neutrality and value-engagement in science

Gerhard Schurz

Institut für Philosophie, Heinrich Heine University Düsseldorf, Werdener Straße 4, 40227 Düsseldorf, Germany

1 Introduction: the program of value neutrality

The value judgment debate constitutes one of the most enduring discussions in the philosophy of science and social philosophy. The first value judgment debate took place in Germany between 1913 and 1917 when Max Weber (1917) defended the value neutrality of the social sciences against the so-called "*Kathedersozialisten*" (professors advocating social policy), who argued that social scientists should proclaim value judgments from the lectern with scientific authority. Weber vigorously countered that objectivity in science can only be achieved if scientists limit themselves to descriptive statements of facts and clearly separate these from their value attitudes, for the reason that value judgments are neither logically nor empirically justifiable but arise from subjective human interests and worldviews.

A second value judgment debate occurred in the German-speaking world during the 1960s and 1970s in the context of the positivism dispute (cf. Albert/Topitsch 1971). Here, proponents of "critical theory" (particularly Habermas in 1968) opposed the ideal of scientific value neutrality, arguing that science is inevitably bound by interests, and the only question is *which* interests scientists serve—a view that was fiercely contested by proponents of the empirical-analytical approach.

A third value judgment debate developed in Anglophone philosophy, starting with papers of Rudner (1953), Jeffrey (1957), and Hempel (1965), during which several novel objections to scientific value freedom were raised, leading many philosophers of science to advocate values in science (e.g., Douglas 2000, 2009, Schroeder 2021, Holman/Wilholt 2022, Elliot 2017). Nevertheless, numerous defenders of the ideal of value freedom can still be found today (e.g., Betz 2017, Lacey 2013, Henschen 2021, Parker 2024).

The advocacy for values in science can be understood in a *weak* and a *strong* sense (cf. Betz 2017, 96f.). Weak in the sense of value engagement, meaning that scientists should strive to make their knowledge useful for the values of its users; and strong in the sense that the justification of scientific knowledge depends on or should depend on non-epistemic value assumptions. My position advocates for values in the weak but not in the strong sense. I will therefore refer to my preferred version as the demand for *value neutrality*, as opposed to a general understanding of "value freedom," which demands that value statements should not appear in science at all. According to the position defended in this paper, value engagement in science is desirable, but a substantive dependence of scientific knowledge on non-epistemic values can and should be avoided. I will argue that a substantive value-boundedness must have a destructive impact on the enterprise of science and its reputation in society, while the properly understood value neutrality of science constitutes an important building block of democracy.

The value neutrality thesis developed here is based on a (at least partially) novel idea of how the value-independence of scientific knowledge can be connected with scientific value engagement: the idea of *hypothetical value statements*. These are conditional value recommendations derived by means-end reasoning from hypothetically assumed values, which do not come from scientists but from science users. The idea of hypothetical value judgments forms a central component of my position, together with a justification of why value independence in science is not only possible but also desirable.

In the next section, the value-neutrality thesis will be cleared from common misunderstandings and precisely defined (Section 2). Then, the idea of hypothetical value statements will be developed and illustrated with experiences from the Covid-19 pandemic (Section 3). Referring to the tragic case of the L'Aquila earthquake, it will be shown how this same hypothetical method can also solve the problem of "inductive risk" without having to resort to categorical value judgments (Section 4). After a brief outline of the metaethical justification of the value neutrality demand (Section 5), further controversial questions will be elegantly resolved through the method of hypothetical value statements, including the "new demarcation problem" (Section 6). Ultimately, it is argued that a value-neutral yet value-engaged position represents the best way to maintain the trust of broad segments of the population in scientific expertise within a democratic society.

2 The requirement of value neutrality: clarification and explication

First, we need a clarification of some important terms. Factual or *descriptive* statements are statements that say something about the factual constitution of the world, including not only to individual facts but also lawlike relations that may be of a strict or statistical sort. Simple value statements have the form "P is (or is not) valuable," where P is a proposition describing a state of affairs or an action. That a state of affairs P is valuable *cannot* be reduced to a factual statement about the value perception of some or many people regarding P (as suggested by Anderson 2004 and Clough 2008), but possesses *normative power*. Therefore, in all ethical systems, value statements and normative statements stand in a close conceptual relation which can be roughly summarized by the formula "The good should be done, and the bad should be avoided" (see Section 3). Value and normative statements are also summarized as *ethical* (or prescriptive) statements. The first prerequisite for fulfilling the requirement of value neutrality is to distinguish between factual statements and value statements in the practice of judgment. In natural language, this is not always easy, but with good will, it is always possible; more on this in Section 5.

In the value judgment debate, a variety of arguments have been introduced that generate important insights but distract from the core question of value neutrality. We will first address four such arguments before we can precisely define the value neutrality thesis.

1.) It has been argued that the social sciences deal with people's value attitudes as the object of their research and therefore must inevitably make value judgments (cf. Strauss 1971). However, as Max Weber (1917, 499–502) already pointed out, it is perfectly possible for the social sciences to empirically investigate the actual value attitudes of individuals or societies without making value judgments themselves. The claim that a person or society holds certain value attitudes or makes value judgments is not itself a value judgment but a descriptive statement, whose confirmation is based on empirical data.

2.) Already in the second value judgment debate it has been pointed out that science cannot be absolutely value-free, because science itself is based on certain so-called science-internal or *epistemic* values—primarily, the value of the pursuit of objective knowledge (Schmidt 1971)—and the same point has been made later in the Anglophone debate (Doppelt 2007). The reference to epistemic values is undoubtedly correct, but nevertheless irrelevant, as value neutrality concerns only non-epistemic values such as wealth, power, prestige, etc.; epistemic values are not what is meant here.

3.) As mentioned at the outset, the demand for value neutrality is often oversimplified into the notion of "value freedom," according to which values "have no place" in empirical sciences. This is a gross mischaracterization, as scientists can attain a number of *derived* values (or norms) from given categorical and fundamental values (or norms) using *means-end reasoning* based on empirical knowledge, and pass these derived values on as *recommendations of means* to the knowledge *users*. These recommendations of means are understood by science in a hypothetical or conditional sense, that is, *relative* to the assumed fundamental values: *If* certain fundamental values are assumed, *then* certain recommendations of means emerge as derived values. A value statement is termed *categorical* if it takes the form "*P* is (or is not) valuable," and it is termed *hypothetical* or conditional if it is an implication between categorical value statements. A categorical value statement is called *fundamental* if it cannot be derived from other value statements through means-end inferences; otherwise, it is termed *derived*.

The identification of suitable *means* for given *ends* is the most important practical task of empirical-descriptive science (see Section 3). What the demand for value neutrality excludes is only that the sciences establish fundamental categorical values (or norms).

4.) The entire process of scientific research is usually divided into three phases (cf. Schurz 2014, sec. 2.1, 2.5.3): In the *context of origination (CO)* (also called "context of discovery"), research questions are first defined. In the *context of justification (CJ)*, data is collected, and hypotheses are generated and tested. In the *context of application (CA)*, well-established findings are finally applied to various purposes. Non-epistemic values assume a role in both the CD and the CA. Which research questions are considered important enough to address is partly influenced by non-epistemic interests, and this is even more true for the choice of ends scientific knowledge is applied to. What the demand for value neutrality addresses alludes exclusively to the CJ.

Summarizing we can explicate our thesis as follows (see Figure 1):

Explication of the Value Neutrality Thesis: The justification of scientific knowledge should be independent of fundamental non-epistemic value assumptions.

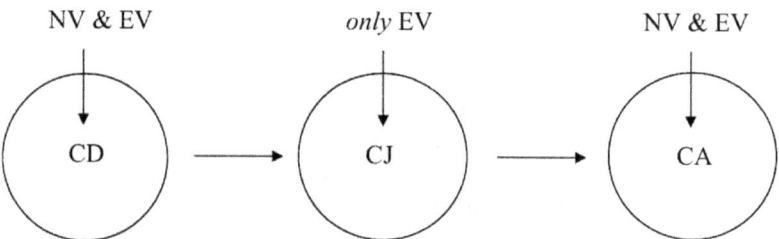

FIGURE 1. Schematic representation of the value neutrality requirement. *Legend:* CD – context of discovery, CJ – context of justification, CA – context of application; NV – non-epistemic values, EV – epistemic values.

The above value neutrality thesis is a *demand*, not a matter of fact, as not all scientists adhere to this standard. The content of the value neutrality demand can, alternatively, also be expressed as the *thesis of value independence*: the *justificatory status* of scientific claims is independent of non-epistemic values. This is a statement about the normative-epistemic status of scientific statements and not about scientists' practices.

Douglas (2000) has emphasized, too, that non-epistemic values influence *several* stages of the scientific research process. However, she refers specifically to the *context of justification*, such as the choice of scientific method. This, of course, is no longer compatible with the value neutrality thesis defended here. Supporters of Douglas (2000) often argue as follows: on the one hand, non-epistemic interests are decisive for the formulation of the *research question* in the CO, and on the other hand, the research question influences the *choice of methodology* in the CJ; this would prove that non-epistemic interests *indirectly* affect the choice of methodology.

While the argument of indirect influence is indeed correct, it nevertheless leads to confusion, because no direct influence is present: *the determination of the research question completely screens off extra-epistemic interests from the choice of the epistemically optimal method.* In other words, extra-epistemic interests may determine *which* research question is investigated, but once the research question is precisely defined, the choice of the optimal method for finding a true or most probable answer is a purely epistemic matter. The concept of *screening-off* goes back to Reichenbach (1956) and is defined within the Markov theory of causality (Spirtes et al. 2000, Sec. 3.4.1–2; Schurz and Gebharter 2016): direct causes screen-off indirect causes from the effects, meaning that if the direct cause of an effect is fixed, knowledge of the indirect cause does not exert any additional influence on the probability of the effect.

Our thesis that the precise formulation of the research question screens-off non-epistemic value influences from the epistemic question of method choice aligns well with Parker's (2024) *theory of epistemic projection.* According to this theory, non-epistemic values are projected onto a research problem, which is then addressed based solely on epistemic criteria, with the answer being more or less useful to the original non-epistemic interests depending on the outcome (ibid., 19, 33). The theory of screening-off explains why answering the research question is a purely epistemic matter, although non-epistemic interests were initially at play. We illustrate this with an example. A research project on efficient treatment methods for *depression* may be driven by the non-epistemic interests of the pharmaceutical industry, which are focused on *pharmaceutical* therapy options: which antidepressants have the best effects (operationalized in terms of efficiency, sustainability, etc.)? This is a precise question, and the choice of the optimal method for this question (data collection, statistical procedures, etc.) *no longer* depends on non-epistemic interests—provided our solution to the "inductive risk" issue in Section 4 is accepted. This holds true even if it turns out that for certain types of depression, *psychological* therapies are more effective than antidepressants. In that case, a rigorously conducted study would yield a largely negative result, meaning antidepressants would have only limited success. Consequently, in the context of justification, it would become clear that restricting research to pharmaceuticals is *counterproductive* for exploring efficient depression treatments. In other words, *even if* non-epistemic motives are initially at play in the CO, their influence can be *corrected* by insight gained in the CJ. The same example also illustrates why the so-called *positivity bias* in research, according to which negative research results are often not published, can lead to significant research distortions (Ioannidis 2005, Schooler 2011). However, this is not an inherent characteristic but a defect of research under career pressure; many medical journals have now implemented measures to ensure the publication of negative research results.

3 Means-end inferences and the significance of hypothetical value statements in science

In its simplest form, the means-end inference—abbreviated as ME inference—has the following form, which can be formulated for both value and norm statements (separated by "/"):

ME (means-end) inference, simple form:

Descriptive means-end premise: Under the given conditions, M is a necessary—or alternatively, an optimal—means for achieving the end E.

Therefore—hypothetical value/norm statement: If E is accepted as a fundamental value/norm, then (the realization of) M is a derived value/norm.

In the ME inference, it is assumed that M is either a *necessary* means—i.e., every end-realizing action entails M—or an *optimal* means, i.e., M maximizes the overall utility with regard to side effects. However, if M is merely a *sufficient* means, then the means-end inference is invalid. This is because for many sufficient means the costs of their side-effects outweigh the benefit of achieving the end. For instance, for letting fresh air into a room breaking through the wall is a sufficient means, too; an optimal means is opening a window and a necessary means is any form of opening to the outside.

The logical form of the inference from ends to necessary means, abbreviated as the NME inference, is straightforward. The descriptive premise "B is a necessary means for end A" can also be rephrased without the terms "means" and "end" through the law-like statement "B is a necessary condition for A," formally "$\Box(A \to B)$" (necessarily, if A then B), where in ethics $\Box$ represents *practical* necessity (meaning, roughly speaking, nomological consequences of boundary conditions, e.g., concerning our planet, that cannot be practically changed). The NME inference then has the following logical form, where O stands for the value operator (OA for "A is valuable") or the obligation operator (OA for "A is obligatory"):

> Logical form of the NME inference:
> Descriptive premise: $\Box(A \to B)$
> ───────────────────────────────
> Hypothetical value/norm conclusion: $OA \to OB$

For better understanding, the following explanations are added:

1.) In $\Box(A \to B)$, B can be a necessary condition for A as well as a necessary consequence of A; the inference holds for both cases. OA represents the fundamental value (or norm), and OB represents the derived value (or norm).

2.) The NME inference can also be transformed into the following inference using the deduction theorem: Premise: $\Box(A \to B)$, 2. Premise: OA / Conclusion: OB. In this form, however, the inference becomes problematic in the sciences, as the second premise is no longer scientifically justifiable. Therefore, if the NME inference is presented in the form above, the second premise must be explicitly *marked* as a value assumption and clearly separated from the first descriptive premise.

3.) The logical form of the inference from the end to *optimal* means is more complex and will be determined further below.

4.) As mentioned, both the ME inference and the connection between values and norms—abbreviated as VN connection—are accepted as *analytically* valid in most ethical theories. By an analytically valid statement, we either mean a logical truth or an extralogical meaning convention (or a statement that logically follows from it). The analytical nature of the VN connection and the NME inference is established in different ways in deontological versus teleological ethical theories (Frankena 1963, 13–16). *Deontological* theories are based on the concept of *intrinsic value*: an action is intrinsically valuable if it obeys certain supreme moral principles, *regardless* of its consequences. Every intrinsically valuable action is obligatory—Kant (1785, BA 52) calls this a "categorical imperative"—and the ME inference is deontologically valid; the obligation of B following from the obligation of A by the ME inference is called a "hypothetical imperative" by Kant (1785, BA 40). In *teleological* theories, by contrast, the value of an action or state of affairs is determined based on the value of its *consequences*. The *overall value* of an action is defined as the "sum" of the utility values of all of its consequences, with benefits being positive and costs being negative utility values. The VN connection principle of teleological theories characterizes an action as obligatory if it maximizes overall utility (or if it is logically entailed by all utility-maximizing actions if there are multiple equally utility-maximizing options; cf. Schurz 1997, Ch. 11.2). The fact that all options for action must be physically or practically possible validates the ME inference for teleological imperatives. Applied to values, the ME inference only holds for *overall values*, not for intrinsic values. The latter would result in the notorious "sanctification of the means by the ends," which is often unacceptable. For instance, it is intrinsically valuable to keep a promise, but if this is only possible at the cost of a human life, then the harm of the action's consequence outweighs the intrinsic value of the action; this results in a negative overall value of the action A, and the NME principle remains valid, i.e., $\Box(A \rightarrow B)$ holds true, but both $\neg \mathrm{W}B$ and $\neg \mathrm{W}A$ also hold.

Based on these specifications, we come to the consequences for understanding value neutrality. As explained, the fundamental purpose of the ME inference is taken over by the expert from the knowledge users, politicians, or practitioners. The expert returns to them a recommendation of means, which must be hypothetically relativized to the assumed fundamental purpose. Only this relativization allows the knowledge users to check whether the fundamental values assumed by the expert are also their own values. The combination of the demand for value neutrality and the ME inference thus accounts for the user's claim to maturity and strengthens democracy. On the other hand, if experts omit the hypothetical value relativization and formulate their recommendations categorically, this can have the politically problematic consequence that knowledge users may be led to actions that do not align with their own interests—which can seriously undermine trust in expert judgments (cf. Elliot et al. 2017; Intemann 2024, 10).

However, the above ME schema is *oversimplified*. The reason a potential knowledge user rejects the expert's assumed ultimate purpose is usually that the costs of the recommended means outweigh its benefits from the user's perspective. The listing and evaluation of the side effects of the recommended means is *omitted* in the above schema. It can have harmful consequences if users are not informed about possible negative side effects. In the case of pharmaceutical drugs, informing about side effects has long become a matter of course, but in expert recommendations concerning the Covid-19 pandemic the duty to inform about side effects was often neglected in expert recommendations (see below). Consequently, we propose refining the above necessary-means-end inference as follows:

Side-effect-transparent schema of the necessary-means-ends (NME) inference:

Descriptive premise: M is, under the given conditions, a necessary means for achieving the end E, and this means has the side effects S.

Therefore—hypothetical value/norm statement: If E is accepted as a fundamental value/norm—and therefore the benefit of realizing E outweighs the costs of S—, then (the realization of) M is a derived value/norm.

The side-effect-transparent version of the NME informs about all the consequences of the recommended means and not just about the intended consequence. Observe the practical dialectics of the side-effect-information between the em dashes: Logically speaking, this information follows already from the premise that E is accepted in the "overall sense". However, most people do not distinguish between intrinsic and overall values. If the information about side effects were omitted, they could be uncritically inclined to agree with value E, which they might not do if they were informed about the side effects. Therefore, adding this information is so important, even if it is logically redundant. Thus, logically speaking, the side-effect-transparent NME inference still has the form $\Box(E \to M)/OE \to OM$, but now with the additional premise-information $\Box(M \to S)$, from which $OE \to OS$ follows, i.e., OE can only hold if the side effects S are acceptable.

Reconstructing the ME inference for *optimal means* is more complicated, as the statement that M_i is "optimal" among all possible means $M_1, \ldots, M_n$ is, strictly speaking, not a descriptive but a normative-evaluative statement and therefore belongs in the conditional part of the hypothetical conclusion. Each means M_i now has its own side-effects S_i. For a *hypothetically* assumed cost-benefit assessment of all consequences, the overall utility of these means must be determined, justifying that M_i is the means with maximum overall utility among and this overall utility is positive:

Side-effect-transparent schema of the optimal-means-ends (OME) inference:

Descriptive premise: $M_1, \ldots, M_n$ are the (practically) possible means for achieving the end E, and the means M_i have the side effects S_i ($1 \leq i \leq m$).

Therefore—hypothetical value/norm statement: If (a) E is accepted as a fundamental value/norm and (b) a cost-benefit assessment is assumed under which M_k has maximal overall utility among $M_1, \ldots, M_n$—and therefore the benefit of E outweighs the costs of the side-effects S_k—then (the realization of) M_k is a derived value/norm.

The content of the premise can be logically expressed by the implication $\Box(E \to (M_1 \vee \ldots \vee M_n))$, i.e., if E is to be attained, one of the possible means has to be realized. The content of condition (b) can be qualitatively represented by the implication $O(M_1 \vee \ldots \vee M_n)) \to OM_k$, i.e., if one of the possible means for E should be realized, then the best one (M_k) should be realized. The condition between the em dashes is again logically redundant and follows from the joint effect of conditions (a) and (b). In conclusion, the above inference can be formalized as follows:

OME inference:

$$\Box(E \to (M_1 \vee \ldots \vee M_n))/(OE \wedge (O(M_1 \vee \ldots \vee M_n) \to OM_k)) \to OM_k$$

In this reconstruction, the OME inference follows from the NME inference, as the premise implies $OE \to O(M_1 \vee \ldots \vee M_n)$, which implies the conclusion.

Different value assumptions can lead to different cost-benefit assessments and thus to different optimal choices of means. This is particularly relevant for democratic societies in a state of strong *value polarization* (cf. Abramowitz und Saunders 2008, Le Bihan 2024, 3). In this situation, it becomes advisable for value-neutral policy advice to conduct hypothetical cost-benefit assessments for *several different* value preferences that reflect the political positions of relevant population groups and are presented to the audience as "alternative options".

The problem of different value weightings became especially pressing during the *COVID-19 crisis*. The value weighting imposed on us by the coronavirus was one between equally fundamental values that came into conflict: health on the one hand, and freedom and well-being on the other. Value decisions of this kind, besides considerations of reason, always depend on actual human interests and therefore contain an inevitably subjective component. Some people were willing to forgo freedoms such as communication, sports, and culture for months for a certain statistical increase in health security, while others found this completely disproportionate. Such subjective differences in attitude must be *acknowledged*, which is why collective value decisions must be tied back to democratic majorities and cannot be dictated by experts, no matter how important expert knowledge may be for understanding the consequences of actions to be evaluated. Therefore, one should expect an expert recommendation in the COVID crisis to name not only the expected benefits but also all costs and to conduct a hypothetical value assessment.

The statement of a commission of the Leopoldina (Germany's National Academy of Sciences) of December 8, 2020, did not meet this value neutrality standard: it stated that a "hard lockdown" was "absolutely necessary from a scientific point of view".[1] This was an is-ought fallacy, as no normative conclusion can be derived from facts (see Section 5). Moreover, the evaluation of the costs of a lockdown compared to the expected benefits was missing, which led to a critical discussion (Wiesing et al. 2021). A year later, the Leopoldina statement of November 27, 2021, had made remarkable progress. It presented two possible options: Option 1: rigid lockdown for everyone, versus Option 2: contact restrictions only for the unvaccinated.[2] The

---

[1] Leopoldina. 7. Ad-hoc-Stellungnahme. Coronavirus-Pandemie: Die Feiertage und den Jahreswechsel für einen harten Lockdown nutzen. 8 December 2020.

[2] Leopoldina. 10. Ad-hoc-Stellungnahme. Coronavirus-Pandemie: Klare und konsequente Maßnahmen – sofort! 27 November 2021.

demand for hypothetical value relativization was thus met, giving those citizens who preferred Option 2 the possibility to reject the Leopoldina's preferred Option 1 *without* resorting to the corner of science denial and "alternative facts."

In summary, we have shown in this section that value neutrality and value commitment, from the perspective of means-end inferences, are not opposites but two sides of the same coin. Why, then, is there persistent talk of the blanket "value-free ideal" instead of value neutrality even in recent debate?[3] Presumably, this has several reasons. As Douglas and Branch (2024) explain, in Western societies in the mid-1950s, a conception of science was predominant according to which the pure sciences should only serve the search for truth and were freed from social responsibility. This conception of science also included the said blanket "value-free ideal." In the 1950s the value-free ideal was further supported by the then widespread view of the *non-cognitivism* of values, according to which value statements were non-rational emotive expressions. However, the alternative view soon gained ground in Analytic Philosophy that a rational treatment of value questions is both possible and socially necessary, and since the 1970s, the disciplines of deontic logic, rational decision theory, and analytic ethics have rapidly established.

Compared to the 1950s, today's scientific self-image has changed significantly. Leading scientific associations now assume that scientists should take partial social responsibility for the foreseeable consequences of their knowledge and therefore engage with its instrumental usability, not only in "applied" but also in "pure" sciences (Douglas/Branch 2024, 12). Although this view perfectly fits our understanding of hypothetical means-end inferences, Douglas and Branch bypass this possible solution and instead argue that the value-free ideal should be rejected and science should be made dependent on categorical values (ibid., 13). From our perspective, this is a clear non sequitur. This brings us to a further reason why the narrow-minded notion of value neutrality as the absence of values lasts so long among opponents of value neutrality: because it is used as an easily criticizable straw man. For example, Dupré (2007) has argued that even if it were possible to separate the factual content from the value content of knowledge, such a separation would be counterproductive because then science could not have practical consequences. The contrary is true, however, as scientific value recommendations are not only allowed but explicitly welcomed by the demand for value neutrality, as long as they appear in the form of hypothetical value statements.

---

[3]Wilholt (2009), Betz (2017), Henschen (2021), Douglas/Branch (2024), Parker (2024), and more.

4 The argument of inductive risk and its resolution: uncertainty transparency and hypothetical cost-benefit assessment

In the third value judgment debate, a novel objection to the thesis of value neutrality was raised which did not play a role in earlier controversies: the argument of inductive risk, or AIR for short. This argument, which goes back to Rudner (1953) and was developed further by Douglas (2000), assumes that in most cases, scientific hypotheses are only supported by empirical evidence with a certain *probability*. When scientists accept a hypothesis H as true, they take a certain 'inductive risk' that H is false: the so-called *error risk*. For example, if the probability of H is 95%, then its error risk is 5%. But what probability of H is still high enough to reasonably accept H as true? For a practically relevant hypothesis the decision to accept it as true means relying on it, that is, being willing to *act* on its basis. Acting on the basis of H brings a benefit if H is true, but a costs if H is false. However, if the costs are much higher than the benefit, then even a hypothesis probability of 95% may not be sufficient to justify acting on the basis of H. Therefore the acceptance of a hypothesis as 'true' inevitably involves extra-epistemic evaluations.

To avoid misunderstandings, we do not claim that all scientific judgments carry a non-negligible risk of error (Betz 201, 98). Judgments such as "The Earth has one moon" (etc.) can be considered practically certain and safely asserted as knowledge categorically. Other hypotheses, like the Big Bang theory, are not practically relevant, so their acceptance is independent of non-epistemic values for this reason. However, many scientific judgments are uncertain *and* practically relevant—and it is to these kinds of judgments that the AIR refers.

A number of authors see the AIR as 'compelling evidence' for the influence of extra-epistemic values on the context of justification.[4] Here, the opposite is to be shown: AIR is much weaker than thought. For AIR can be refuted by the demand for *uncertainty transparency*—i.e., the explicit indication of the involved error risks—and this demand not only serves value neutrality but also prevents questionable consequences of expertocracy.

To refute AIR, we use the prediction of *earthquake safety* as an example. Let H be the hypothesis that (in a given area in the next few days) *no* earthquake of a Richter-scale-magnitude greater than 6 will occur. If H is accepted, then the affected people stay in their homes, which, if H is true, brings a comparative benefit of zero (no additional costs), but if H is false, it brings very high costs C_h, which can mean injuries or death, at least for those whose houses are at risk of collapsing at this earthquake magnitude.

---

[4] See Wilholt (2009), Steele (2012), Douglas/Branch (2024); Holman/Wilholt (2022) even speak of a "general consensus".

On the other hand, if the affected people accept non-H, considering an earthquake likely, then the evacuation of their homes is carried out, which brings comparatively lower costs C_l, both if non-H is true and if non-H is false. If p denotes the probability of an earthquake, i.e., the counter-probability of H, then the expected utility of H, abbreviated as $E(H)$, as well as that of non-H, are as follows:

$$E(H) = (1-p) \cdot 0 - p \cdot C_h = -p \cdot C_h.$$
$$E(\text{non-}H) = -(1-p) \cdot C_l - p \cdot C_l = -C_l.$$

Accepting H is better than rejecting it, if and only if $E(H) > E(\text{non-}H)$, i.e., iff $C_l/C_h > p$.

The acceptance of the hypothesis H (no earthquake) is therefore only reasonable as long as the ratio of the costs of its erroneous acceptance to the costs of its erroneous rejection is greater than its error probability p—in our example, greater than 5/100.

It is undoubtedly true that the question of accepting non-H versus H in a qualitative sense—and thus the question of whether to evacuate or not—depends on extra-epistemic cost-benefit considerations. But does this really mean that scientists should make such decisions on behalf of the affected people, as proponents of AIR would suggest? My answer to this question is: *No*, this is actually often the worst thing scientists can do in such a situation. Instead, scientists should explicitly state the involved error risks and make them transparent to non-experts through *hypothetical* cost-benefit considerations. We call this the condition of *uncertainty transparency*.

However, if scientists present qualitative statements *categorically* and conceal the error risk, unpleasant consequences may easily ensue for both the knowledge users and the knowledge producers. The earthquake example illustrates this clearly. In October 2012, six earthquake experts and a government official in Italy were sentenced in the first instance to several years in prison because they did not predict an earthquake that occurred on April 6, 2009, in L'Aquila, in which more than 300 people died, but instead gave the all-clear.[5] The case attracted worldwide attention. *If* the judges' view were correct that the earthquake experts gave the all-clear categorically at that time, despite a slightly increased risk due to the registered tremors, then the earthquake researchers would indeed bear partial responsibility: because they acted in the sense of Rudner and Douglas's recommendation, took the cost-benefit assessment out of the hands of the affected individuals by announcing an "all-clear", and thereby concealed the involved risk. Even if the risk of an earthquake of magnitude 6 to 7 increases from only 0.5 to

---

[5] For the following, see Edwin Cartlidge: "Seven-year legal saga ends as Italian official is cleared of manslaughter in earthquake trial. Verdict follows conviction of deputy for advice given ahead of L'Aquila earthquake." *Science*, 3 October 2016.

1%, this may be reason enough for one person to endure the hardship of leaving their home due to their specific circumstances (for example a mother with her kids), while another person is willing to take the risk (for example a farmer who cares for his livestock). Which scientist would wish to take the decision out of the hands of the affected individuals in such a case, citing their "expert authority", and be held accountable for the consequences under threat of punishment? Probably no one.

The earthquake researchers defended themselves against the court's judgment by pointing out that they did not issue a categorical all-clear, but rather a more cautious statement, stating that the registered slight increase in tremors was not significant but still "within the normal range." In 2014, the judgment was revised, and the involved scientists were acquitted; only the government official, who had conveyed the "all-clear" message to public media, was convicted. In any case, this remarkable story clearly demonstrates that categorical evaluations of risk consequences are *entirely* outside the purview of scientists.

The counter-argument against AIR, which asserts that uncertain scientific statements should be framed as *probabilistic statements*, was first articulated by Jeffrey (1956, 237). More recently, referencing Schurz (2013), Betz (2017) and Henschen (2021) have advocated addressing the problem of inductive risk by making error risks explicit. Douglas and like-minded authors counter this position by arguing that the public and their political representatives desire definitive statements with clear action implications, such as "all-clear," from the "authority of science" (Douglas 2000, 563; 2009, 135; John 2015, 82; Wilholt 2009, 94). However, scientists must not yield to such false expectations. Rather, it is their duty to educate people about the limits of what is knowable—which in this case means informing the public about the error risks of their predictions. Nonetheless, we agree with the authors that merely providing a probabilistic significance statement, hardly understandable for laypeople, is insufficient for such public enlightenment. Besides indicating the error probability of scientific predictions or assertions, it is crucial to make this error probability comprehensible to non-experts. Here again, our proposal relies on the idea of *hypothetical action recommendations based hypothetical cost-benefit evaluations*. In our earthquake example, this would mean illustrating the practical significance of a 5% error probability as follows: *If* the potential negative consequences of an earthquake during one's stay at home are assessed to be more than 20 times higher than the costs of a temporary evacuation, *then* it is advisable to undertake the evacuation. This proposal goes far beyond Jeffrey's requirement that experts should simply report the error the probabilities: it takes justice to Douglas (2020) requirement that experts cannot merely report error probabilities and leave the decisions to others, but due to the hypothetical mode, this now takes place in a value-independent manner.

After this fundamental defense of value neutrality against the AIR, three *refinements* of our argumentation are presented in the context of the current debate.

1.) Rudner (1953, 4) and Douglas (2009, 53f.) developed a fundamental objection to the proposal that scientists should limit themselves to probabilistic statements. They argued that even if scientists only assert a probability, they have already accepted a hypothesis, namely a probability hypothesis, which itself is uncertain, thereby perpetuating the problem. If this were true, the probabilistic proposal would be subject to the problem of *infinite regress* and would collapse. But closer inspection reveals that this is not the case. Suppose there is second-order uncertainty about the given error probability of the following form: "With 95% probability, the error risk of H lies between 4% and 6%." This is sometimes assumed in Bayesianism, but by forming expectation values or confidence intervals one projects the second level information back to the first level. The only practical question affected by a second-order evaluation is whether to base the decision on the probabilistic *expectation value* of the risk (which in our example is 5%) or to quantify the risk using the lower and upper risk-limits of 4% and 6% (according to the confidence interval method; see Cox and Hinkley 1974, 49, 209). The second variant weakens the decision rule's discriminatory power and increases the possibility of stalemates. Otherwise, there are no practically relevant consequences, and at the next-higher (third) evaluation level, the practical consequences are zero. Therefore, the regress stops at the second level at the latest, and the regress argument is refuted.

2.) Douglas (2000, 563) argues in her contribution that the influence of non-epistemic values cannot be avoided by limiting statements to probabilities because an error risk arises not only in the formulation of qualitative statements but also at other points during the internal research process, particularly in *data description* (ibid., 569f.) and the *interpretation of results* (ibid., 573f.). This is true, but I don't see a fundamental problem here. All these error risks add up to the overall risk of the final hypothesis H, according to the laws of probability theory. To decide whether H should be accepted as true and used as a basis for action, it is only necessary to know this overall risk. One may object that some of these risks rely on implicit knowledge and are not reflectively available to the scientists. This may be true, but recall that value neutrality is a normative requirement: it requires that these risks *should* be made reflectively available as far as possible, and even if they are entirely unknown or based on shaky guesses, this has to be stated instead of being suppressed.

Moreover, the examples Douglas cites do not solely concern error risks but also inadequate methodological steps. Concerning data description, Douglas addresses the question whether borderline cases between non-carcinogenic and carcinogenic lesions in lab rats should be classified as carcinogenic or non-carcinogenic. She argues that this question is influenced by extra-epistemic risk assessments (ibid., 571f.). However, both options are epistemically inadequate, since non-assignable borderline cases should be marked and excluded from statistical calculations. Concerning the interpretation of results, Douglas discusses whether a found correlation between dioxin and increased cancer rates should be interpreted by a continuous increase model or a threshold model. This is also not a matter of extra-epistemic preferences but can be answered through refined empirical methods such as nonlinear curve regressions.

3.) In the discussion of AIR, the tolerable probability of the α-and β-error in the context of choosing the *significance level* is frequently referenced (cf. Wilholt 2009, Section 6; Henschen 2021; Douglas 2000, 563f.). We conclude this section by addressing this connection. Let H be a statistical correlation hypothesis: the influence of dioxin on cancer incidence in lab rats, from Douglas's example. Testing H involves comparing the cancer frequency in two samples: an *experimental group* of lab rats administered dioxin and a *control group* without dioxin. Following Henschen's (2021, 9–10) continuation of this example, assume that the cancer frequency in the experimental group is 25%, and 15% in the control group. A statistical significance test then asks: is the frequency increase by 10% due to chance, or is it "significant," i.e., attributable to a presumed causal relationship? The α-error denotes acceptance of H when H is false (or rejection of the opposing "null hypothesis," $\neg H$, when the latter is true). The probability of the α-error is the likelihood that a difference Δ at least as large as the *observed difference* $\Delta_o = 10\%$ would be randomly found between two samples. (In our example, we only consider positive differences, i.e., a 'one-sided t-test' is conducted.)

The probability distribution of the frequency differences Δ between two 100-element random samples of a binary characteristic is a normal distribution; this is depicted in Figure 2 by the solid left (blue) curve. Using the underlying distribution formula (Henschen 2021, 9), one calculates the probability of randomly observing a difference of at least Δ_o to be approximately 4%. The number 4% is the so-called "p-value" of Δ_o respectively the error probability of H—in Figure 2, this corresponds to the proportion of the area below the normal distribution curve at the right side of the value 10.

Note: strictly speaking, p is not the error probability of the hypothesis but rather that of the underlying test-statistical *procedure*—i.e., the probability of observing a frequency difference of at least Δ_o in pairs of 100-element random samples amounts to 4% (Cox/Hinkley 1974, 49, 209). Only under the assumption that the data were *representative* for H is it legitimate to interpret p as the error probability of H given Δ_o.

In test statistics, the significance level refers to a probability threshold s that the p-value of the observed difference must fall below for H to be acceptable as "significant"; this threshold corresponds to a "minimum significant difference." Typically, s is set at 5%; in our example with sample sizes of 100, the minimum significant difference is calculated to be 9%, which is indicated by the vertical line in Figure 2. This definition is made *pragmatically* in test statistics and is by no means mandatory. Rather, a categorical definition of certain "acceptance thresholds" has to be criticized, because as explained, the scope of acceptable error probabilities depends on external cost-benefit evaluations. In fact, a categorical setting of acceptance thresholds does not correspond to standard statistical practice; instead, the mentioned p-value of the observed difference is always explicitly indicated as the correlation hypothesis' error risk.

The significant sample difference decreases proportionally to the increase of the square root of the sample size $n(\Delta \sim 1/\sqrt{n})$. This is illustrated in Figure 2 by the dashed left (blue) curve, representing the steep-peaked normal distribution for a sample size of 2000. Any minimal sample difference becomes significant if the samples are sufficiently large. Therefore, the mere assertion that a "highly significant" relation was found between two variables is a weak claim without specifying the sample size, and merely indicates some possibly *very slight* correlation (here between dioxin and cancer rate). Far more important is the indication of the so-called *effect size*, i.e., the strength of the correlation, which in our example can be measured by the shift in frequency and amounts to 10% (cf. Andrade 2020).

This brings us to the second error type, the β-*error*, which, complementary to the α-error, denotes the rejection of a true H (thus the acceptance of a false $\neg H$). The probability of a β-error depends not only on the chosen significance level s for the α-error and the sample size but also on the claimed effect size of H. Suppose we are interested in dioxin-induced frequency increases of at least 15%. The probability distribution of the differences between two samples from different populations with a frequency difference of 15% is depicted in Figure 2 by the solid right (red) curve. Assume s is set at 5%, i.e., the minimal significant difference is 9%. Then the probability of a β-error is the probability that a sample drawn from a dioxin-exposed population with a cancer frequency of at least $(15+15=)$ 30% will randomly deviate so far to the left that it still falls within the acceptance interval of $\neg H$, thus leading to the rejection of H. In our example, the probability of the β-error is calculated as 15%; in Figure 2 this is the proportion of the area

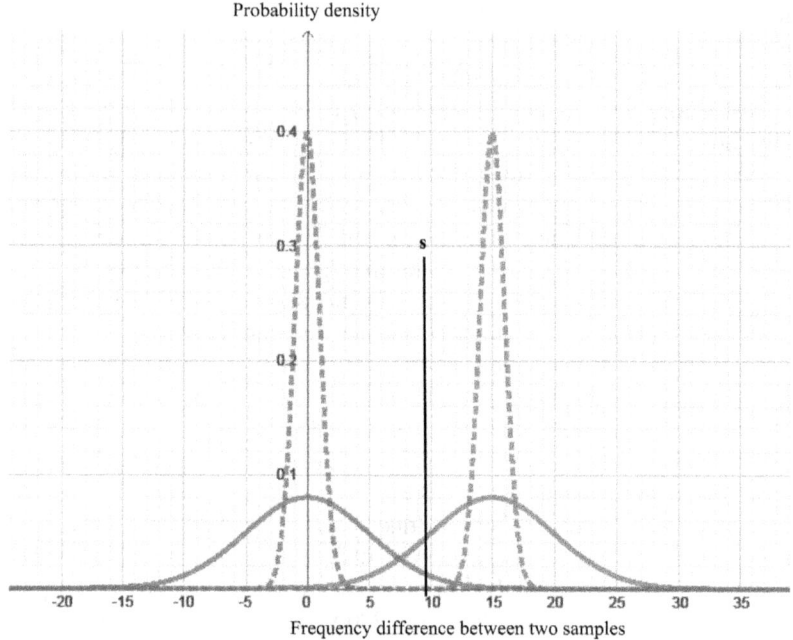

FIGURE 2. Normal distribution of frequency difference between two samples. Left (blue): Assuming the null hypothesis. Right (red): Assuming the alternative hypothesis of a frequency increase of 15%. Solid line: Sample size 100. Dashed line: Sample size 2000. Vertical line: significant difference for significance coefficient of $s = 5\%$ and sample size of 100. α-error: Proportion of area below the left-hand curve to the right of s. β-error: Proportion of area below the right-hand curve to the left of s. (Created with Geogebra.)

below the right solid normal distribution at the left side of the significance threshold s.

For constant sample size and effect size, the probability of a β-error is the greater the smaller the tolerable probability of an α-error. From this, Douglas (2000, 563f) and others have concluded that reducing the α-error risk "inevitably" comes at the cost of increasing the β-error risk (Betz 2017, 95; Parker 2024, 21f.; Wilholt 2009, 94f.). For Douglas, this is the gateway for the influence of extra-epistemic values: a low α-error risk reduces the number of falsely diagnosed dioxin-cancer correlations, thus leading to "underregulation" in the interest of the industry, while reducing the β-error risk decreases the number of overlooked true dioxin-cancer correlations, thus leading to overregulation in the interest of the health-endangered population.

The fact that the α-error risk is kept low in statistics supposedly serves the interests of the industry. However, this argument is misleading for two reasons.

Firstly, by increasing the sample size, it is always possible to circumvent the α-β tradeoff. This is illustrated in Figure 2 by the two dashed curves (left and right) for a sample size of 2000. Since the two steep-peaked distributions barely overlap, this allows the α- and β-error risks to be *simultaneously* reduced far below 0.01%.

Secondly, even when dealing with small sample sizes, the demand for uncertainty transparency requires reporting not only those sample differences that were significant at $s = 5\%$; rather, *all* found sample differences should be reported, including the non-significant ones. For each found difference Δ_o, (a) the error risk of the correlation hypothesis (α-error risk) should be specified, and (b) the error risk of the null hypothesis (β-error risk) for an assumed minimum effect size, *if* this null hypothesis were claimed at the found difference Δ_o (i.e., if s were identified with the p-value of Δ_o). Step (a) is standard practice in statistical investigations. Step (b) is often omitted but would be helpful in providing transparent information for statements with higher α-error risk and lower β-error risk, in line with Douglas. Illustrated in terms of our example: suppose we find a frequency increase of only $\Delta_o = 5\%$. Then the error probability of H given $\Delta_o = 5\%$ is calculated to be 19%, and the error probability of $\neg H$ for effect sizes of at least 15% is 4%. Summarizing, clarification of all error risks is possible even in the more complex case of statistical comparison tests, without the need for categorical valuations to be involved.

5 The metaethical justification of value neutrality

Besides its foundation in the philosophy of science, the thesis of value neutrality also rests on certain *metaethical* premises, the thorough justification of which would exceed the scope of this work; however, they should at least be mentioned in this section. The metaethical justification of the thesis of value neutrality is based on two theses:

Thesis 1 – The is-ought or fact-value dichotomy: From purely descriptive statements, value or normative statements cannot be logically or analytically derived without implicitly presupposing ethical premises.

Thesis 1 goes back to David Hume (1739/40, 177f.) concerning logic and to G. E. Moore (1903, 15f.) concerning analytical meaning postulates. However, this alone is not sufficient to justify the thesis of value neutrality. For one could hold the position that there are also so-called *value sciences* that can justify categorical ethical statements with similar objectivity as factual statements are justified by empirical sciences. Those who take this position can agree with the fact-value dichotomy and still argue that fundamental extra-epistemic values have their place in science, as they obtain their objective justification in the value sciences. Several intellectual movements have held such a position; for example, (neo)marxism (Habermas 1968) or value platonism (Scheler 1913/16). Therefore, to justify value neutrality the following second thesis is additionally required:

Thesis 2 – the logical-empirical nature of the scientific method: Scientific knowledge is based on experience and logical-mathematical knowledge. Only through this type of knowledge can objective truth be achieved reliably. Value statements, on the other hand, are dependent on subjective interests or intuitions. Therefore, in the realm of values, no objectivity comparable to that of empirical sciences can be achieved.

Ad 1 – Justifying the dichotomy: Attempts to prove the is-ought dichotomy have led to a number of problems. This has led several philosophers (Searle 1969, MacIntyre 1981, Putnam 2002) to question the dichotomy thesis, which has only recently been placed on a solid logical foundation through newer investigations (Pigden 1989, Schurz 1997, Russell 2022). Three main problems arise in this context:

1.1 The conceptual fact-value distinction: The thesis of value neutrality first requires the separation of descriptive from ethical statements. For formal languages, this distinction is ensured by the different forms of expressions, but for natural language, a problem arises. Many everyday language concepts simultaneously possess factual and value content, often closely intertwined, making the conceptual separation difficult. Some critics of value neutrality, such as Putnam (2002, Ch. 2) and Dupré (2007), have claimed that for so-called "thick concepts" like "rape," a fact-value distinction is impossible. However, as Hare (1981, 74ff.) pointed out, this separation is more a matter of will than of capability and is always possible with the aid of appropriate linguistic means (see Schurz 2013, Sec. 6.2). We illustrate this with Dupré's example of the concept of rape of a woman. The descriptive part of the concept would, at first approximation, be given when a man forces a woman against her will to engage in sexual intercourse or similar acts. The conjunction of the above description and the value judgment "And this is a serious offense" can be seen as a satisfactory approximation of the descriptive-normative overall meaning of "rape," which complies with the separation requirement. The separation of the descriptive and prescriptive content is also necessary for the empirical investigation of a question that Dupré also considers to be of utmost importance, namely which measures can most effectively reduce the frequency of attempted rapes (Schurz 2013, Section 6.2).

1.2 The logical dichotomy: To justify this dichotomy one must show that purely descriptive premises do not logically entail ethical conclusions. The main problem for proving this thesis was that there are also mixed conclusions containing both descriptive and normative components, leading to the notorious "paradox of Prior" (Prior 1960). This problem was solved by applying a relevance criterion, according to which purely descriptive premises never logically entail conclusions containing *relevant* (i.e., not arbitrarily replaceable) ethical components. For example, in the logically valid inference from "Grass is green" to "Grass is green or Trump should lose the election," the disjunctive conclusion component "Trump should lose the election" is irrelevant and can be replaced by anything else without affecting validity (salva validitate). In Schurz (1997), the non-derivability of relevant ethical conclusions from descriptive premises was called the *general Hume thesis* and was proven for all multimodal logics whose axiom schemata do not contain so-called *bridge principles*, or BPs (ibid., Theorems 1, 2, 3, 7, and 8). BPs are sentence schemata that establish a connection between descriptive and ethical statements by containing the same schematic letter both inside and outside the scope of ethical operators.

1.3 The analytical dichotomy: Thirdly, it must be shown that BPs cannot be considered analytically valid meaning postulates and thus cannot be axioms for extended modal logics (Schurz 1997, Ch. 11; 2010, Sec. 3). Moore (1903, 15ff.) justified this thesis with his *open question* argument: for any descriptive condition D it can always be meaningfully asked, "but is D also good?" This shows that the good cannot be analytically reduced to a descriptive condition. Rather, the acceptance of a (substantial) BP is always linked to a certain (usually controversial) ethical position. For example, the BP of utilitarianism states that good is what maximizes the overall utility for all people (Mill 1863). This BP is criticized by alternative ethics for there being other moral intuitions not serving the maximization of utility.

However, the above applies only to so-called *substantial* BPs—these are BPs that can justify categorical value or norm statements. In contrast, so-called *functional* BPs, like the *means-end principle* discussed in Sec. 3, are analytical in nature.[6] However, with functional BPs it is provably impossible to derive *non-trivial* categorical ethical statements from descriptive premises;[7] only *hypothetical* ethical statements are derivable. This result is called the *practical Hume thesis* in Schurz (1997) and is proven for all multimodal systems axiomatized by standard axioms together with functional BPs (ibid., theorem 6, corollary 10). The practical Hume thesis is *fundamental*

---

[6] Logically, a functional BP is characterized by the property that it becomes a logical truth when all ethical operators are removed from it.

[7] A categorical normative conclusion from descriptive premises D is called non-trivial if D does not already entail that the prescribed (or prohibited) state of affairs is always true (or false).

to our value neutrality thesis, as it guarantees the logical possibility of the coexistence of value neutrality and means-end inferences.[8]

Ad 2 – Justifying the logical-empirical nature of the scientific method: Can there be an "objective value science" that can justify categorical ethical statements in the same objective way as empirical science justifies factual statements? Can there be moral experts who can inform people about the correct values as reliably as astronomers inform us about the stars? For this to be the case, there would have to be both substantive and intersubjectively universally valid ethical principles. However, this is doubtful. Based on a classification of ethical theories, Schurz (1997, ch. 11; 2010, Sec. 7) argues that the highest justificatory principles of these theories are either almost *empty* concerning their applications, or they are substantive but highly *controversial*. We can only summarize the most important results of this investigation here:

Utilitarianism: This position is "reductionist," meaning that the good is reduced to descriptive conditions. The fundamental principle of utilitarianism characterizes an action as good if it contributes to maximizing the overall utility for all people. However, this overall utility depends on people's empirical interests, which are culturally dependent and can change. Furthermore, the overall utility depends on the method of utility aggregation—particularly how conflicting utility values are weighed against each other, such as health versus freedom and social welfare in the example of the COVID-19 pandemic. As noted in Section 3, such trade-offs inevitably involve subjective factors.

Theories of Justice: These theories are "autonomist", meaning that they do not base the concept of the good on empirical conditions but on intuitive standards of justice. Here, too, there are different and conflicting principles of justice, particularly the principles of merit-based justice and distributive justice (see, for example, Nozick 1974 versus Rawls 1971). According to merit-based justice, distribution of wealth is considered fair if individuals have acquired their claims to property in a legitimate and equal-opportunity manner; thus, disproportionate wealth acquired through higher performance is regarded as fair. In contrast, distributive justice considers wealth, even if it was earned based on merit, to be unfair and subject to redistribution. The balancing of these two principles of justice is tied to subjective interests or intuitions and varies significantly between the conservative-liberal camp and the leftist-socialist camp.

Intuitionist-empiricist ethics: These ethical theories assume an inherent "moral sense" in humans, which relies primarily on intuitions (Hume 1939/40, 177f.; Firth 1952). The fundamental principle of these ethics characterizes a

---

[8]The result holds, more generally, for all conclusions of the form $Op(D \rightarrow N)$ where N is a categorial ethical formula, D a possibly empty descriptive formula and Op a possibly empty sequence of quantifiers and modal operators.

state of affairs P as morally valuable if every person under normal conditions would spontaneously perceive or judge P as morally valuable. While this principle is analytically plausible, its problem lies in the fact that its scope of application is almost empty. In contrast to genuine perceptions, there are hardly any moral value judgments that are intersubjectively stable across cultures. This is convincingly confirmed for the present by the cross-cultural studies of the World Value Survey by Ronald Inglehart.[9]

In summary, fundamental categorical value judgments cannot be justified with the same claim to scientific objectivity as scientific knowledge. And if scientists occasionally make a fundamental categorical value judgment, they should, in any case, mark it as such and not present it as quasi-scientific knowledge.

6 Consequences for the debate on values in science

In this essay, we have established a form of the value neutrality thesis that differs from traditional formulations of the ideal of value freedom. We aimed to show that a harmonious coexistence of scientific value engagement and value neutrality is possible by formulating value statements hypothetically and justifying them through empirically supported means-end inferences. The method of hypothetical formulation also enabled a solution to the problem of inductive risk by disclosing error probabilities and explaining their practical implications through hypothetical cost-benefit evaluations. In this concluding section, two further issues of the recent value judgment debate are addressed through the hypothetical method.

The first of these issues is the so-called *new demarcation problem*, introduced into the debate by Wilholt (Wilholt 2009, Holman/Wilholt 2022). This is a challenge for all opponents of value neutrality. As numerous meta-studies show, many studies funded by the pharmaceutical industry suffer from a *bias* that benefits the revenues of their sponsors (and often includes the omission of negative results)[10]. The new demarcation problem now consists in the fact that proponents of scientific value dependence criticize such studies due to their capitalist value bias, while *simultaneously* considering a value bias in scientific knowledge as inherently unavoidable and even beneficial. Value proponents should, therefore, have no objections to a capitalism-friendly value bias and should regard it as equally legitimate as their own typically left-leaning value bias (see also Henschen 2021, 20). How can value proponents avoid this consequence? For example, Wilholt agrees with Douglas's thesis that, because of the AIR, extra-epistemic values inevitably influence the justification of scientific statements, but at the same time, he claims that scientists themselves establish *conventional standards*

---

[9]Inglehart (1997) and the website of the World Values Survey.
[10]See Wilholt (2009, 93f.), vom Saal and Hughes (2005), and Brown (2008, 191).

to distinguish legitimate from illegitimate value influences (2009, 96–99). However, Wilholt leaves us unclear about the justification of these standards, so it is not obvious why religious creationists or proponents of capitalism could not also establish standards that serve their own perspectives.

Some authors have suggested that the new demarcation problem could be solved by scientists relying on certain preferred values—namely, *democratic* values that form the foundation of our society.[11] Hilligardt (2023), however, argues that this would exclude "partisan science," which represents the values of socially underrepresented groups, such as *feminist* values. Instead, Hilligardt advocates a *pluralism* of political purposes within the sciences. A more grave objection to the democratic value dependence thesis is the *polarization objection* raised by Le Bihan (2024). Modern democratic societies are *polarized*: the politically right-leaning and left-leaning segments of the population hold opposing value preferences on many issues. The suggestion to rely on democratic values does little to address this issue, as it leaves open which side of the political spectrum these "democratic values" should be on. Upon closer examination, this suggestion even has frightening implications. For it would mean that if a political shift occurs in a democratic society, scientific knowledge and textbooks would need to be *rewritten*, as they depend on the "prevailing democratic values," which may have changed significantly at that moment. For example, if there were a shift in the U.S. from Democrats to Republicans, wouldn't this position suggest to close or reassign certain professorships and potentially entire departments? I doubt that opponents of value neutrality have such outrageous consequences in mind with their arguments. History has seen plenty of unwelcome examples of this kind. Fortunately, our society's understanding of science has largely freed itself from such political constraints, and it should remain that way.

Moreover, empirical studies show that the average political attitudes of university academics are significantly more left-leaning compared to those of non-academic populations (Duarte et al. 2015). Therefore, scientists are even more urged to frame recommendations based on their own values hypothetically rather than categorically, to avoid the risk of losing the trust of the non-academic public. This leads us to a second point of discussion in the recent controversy: the issue of *trust in science*. There is consensus that in our information society, public trust in science is fundamentally important. Proponents of scientific value influences have raised the question of whether transparency regarding these value influences would be beneficial or detrimental to public trust in science (Intemann 2024, 3). While some philosophers of science provide an affirmative answer (e.g., Elliott 2017, NAS 2018, Intemann 2024), others respond more skeptically. Some value-friendly philosophers of science (e.g., John 2018) have even argued that it is better to

---

[11] E.g., Schroeder (2021), Alexandrova/Fabian (2022), and Elliott (2017).

leave non-experts, who "naively" believe in the value-free nature of scientific knowledge, in their false belief, since revealing its actual value-dependence would undermine their trust in science. From our perspective, this argument is *doubly misguided*: first, the understanding of science held by these "naive non-experts" seems closer to the truth than that of their "intellectual critics"; and second, it is cynical to propose that the public should be kept in a false belief in order to better control it. In earlier times, religious leaders who were never really concerned with God but only with power argued in similar ways.

Intemann (2024, 8–10) also argues that it is ultimately a fraud when scientists hide their value assumptions from the public and present their findings as value-free knowledge. We agree with Intemann, but further emphasize that hypothetical formulations of recommendations are likely to best promote trust in science, whereas categorical recommendations are more likely to undermine this trust, especially when they oppose public interests. This has been demonstrated, among other instances, during the Covid-19 pandemic. Studies by Angeli et al. (2021) show that many non-experts lost trust in Covid experts when they realized that their recommendations were based on one-sided value preferences, such as preventing hospital overcrowding at the expense of personal freedoms. In Germany, trust in science decreased from 2022 to 2023 by 25% in the lower and middle education levels.[12] Research by Elliott et al. (2017, 13) indicates that the disclosure of values by experts either increases or decreases public trust, depending on whether the public shares those values. This is a dilemma that can only be resolved by the method of hypothetical recommendations, which protects experts from value assignments and ideological assumptions without depriving them of their usefulness to people.

In conclusion, I attempted to demonstrate that tying scientific knowledge to extra-epistemic values, as advocated by opponents of value neutrality, is not the right approach to integrating value engagement into science; rather, it ultimately leads to a dead end. The only way I see to combine the widely desired objectivity of science with value engagement and social responsibility is through the proposed method of hypothetical value statements and their justification through scientific means-ends inferences. Instead of adhering to extra-epistemic values, scientists should take *pride* in the objectivity and impartiality of their findings. At their core, these findings are neither capitalist nor communist, patriarchal nor feminist, but merely more or less true or probable. Only value-neutral truth orientation allows for stable scientific progress through the various stages of human cultural development.

---

[12]Wissenschaft im Dialog GmbH, Wissenschaftsbarometer 2023, Berlin 2023.

Acknowledgement. For valuable comments I am indebted to Wolfgang Spohn, Martin Carrier, Torsten Wilholt, Paul Hoyningen-Huene, Charles Pigden, Tobias Henschen, Christian Feldbacher-Escamilla and Alexander Christian.

References

Abramowitz, A. und Saunders, K. (2008): "Is Polarization a Myth?", *The Journal of Politics* 70, 542–555.

Albert, H., and Topitsch, E. (eds, 1971): *Werturteilsstreit*, Wissenschaftliche Buchgesellschaft.

Alexandrova, A., and Fabian, M. (2022): "Democratising Measurement: Or Why Thick Concepts Call for Coproduction", *European Journal of Philosophy of Science* 12: 7.

Anderson, E. (2004): "Uses of Value Judgments in Science: A General Argument, with Lessons from a Case Study of Feminist Research on Divorce", *Hypatia* 19, 1–24.

Andrade, C. (2020): "Mean Difference, Standardized Mean Difference (SMD), and Their Use in Meta-Analysis", *The Journal of Clinical Psychiatry* 81(5): 20f13681.

Angeli, F., Camporesi, S. and Dal Fabbro, G. (2021): "The COVID-19 Wicked Problem in Public Health Ethics: Conflicting Evidence, or Incommensurable Values?", *Humanities and Social Sciences Communications* 8, 1–8.

Betz, G. (2017): "Why the Argument from Inductive Risk Doesn't Justify Incorporating Non-epistemic Values in Scientific Reasoning", in: K. Elliott und D. Steel (eds.), *Current Controversies in Values and Science*, Routledge, 94–110.

Brown, J. (2008): "The Community of Science", in: M. Carrier, D. Howard, and J. Kourany (eds., 2008), *The Challenge of the Social and the Pressure of Practice*, University of Pittsburgh Press, 189–216.

Clough, S. (2008): "Solomon's Empirical/Non-Empirical Distinction and the Proper Place of Values in Science", *Perspectives on Science* 16, 265–279.

Cox, D. and Hinkley D. (1974): *Theoretical Statistics*, Chapman & Hall, 49, 209.

Doppelt, G. (2007): "The Value Ladenness of Scientific Knowledge", in: Kincaid et al. (2007), 188–218.

Douglas, H. (2000): "Inductive Risk and Values in Science", *Philosophy of Science* 67, 559–579.

Douglas, H. and Branch, T.Y. (2024): "The Social Contract for Science and the Value-free Ideal", *Synthese* 203, 40.

Douglas, H. (2009): *Science, Policy, and the Value-free Ideal*, University of Pittsburgh Press.

Duarte, J. L. et al. (2015): "Political Diversity Will Improve Social Psychological Science", *Behavioral and Brain Sciences*, 38, e130.

Dupré, J. (2007): "Fact and Value", in: Kincaid et al. (2007), 28–41.

Elliott, K. (2017): *A Tapestry of Values. An Introduction to Values in Science*, Oxford University Press.

Elliott, K. et al. (2017): "Values in Environmental Research: Citizens' Views of Scientists Who Acknowledge Values", *Plos One* 12, e0186049 (doi: 10.1371/journal.pone.0186049).

Firth, R. (1952): "Ethical Absolutism and the Ideal Observer", *Philosophy and Phenomenological Research* XII/3, 317–345.

Frankena, W. (1963): *Ethics*, Prentice Hall.

Habermas, J. (1968): "Erkenntnis und Interesse", Engl. transl. in J. Habermas, *Knowledge and Human Interests*, Polity Press 1987, Appendix.

Hare, R. (1981): *Moral Thinking*, Clarendon Press.

Hempel, Carl G. (1965): "Science and Human Values", in: Carl G. Hempel: *Aspects of Scientifc Explanation*, Free Press, New York, pp. 81–96.

Henschen, T. (2021): "How Strong is the Argument From Inductive Risk?", *European Journal for Philosophy of Science* 11: 92

Hilligardt, H. (2023): "Partisan Science and the Democratic Legitimacy Ideal", *Synthese* 202: 135.

Holman, B., and Wilholt, T. (2022): "The New Demarcation Problem", *Studies in History and Philosophy of Science* 91, 211–220.

Hume, D. (1739/40): *A Treatise of Human Nature. Vol. II/Book III: Of Morals*, Clarendon 1896.

Inglehart, R. (1997): *Modernization and Postmodernization: Cultural, Economic, and Political Change in 43 Societies*, Princeton University Press, Princeton.

Intemann, K. (2024): "Value Transparency and Promoting Warranted Trust in Science Communication", *Synthese* 203: 42.

Ioannidis, J. (2005): "Why Most Published Research Findings are False", *PLoS Medicine* 2(8), e124, 696–701.

Jeffrey, R. (1956). "Valuation and Acceptance of Scientific Hypotheses", *Philosophy of Science* 22, 337–346.

John, S. (2015): "Inductive Risk and the Contexts of Communication", *Synthese* 192, 79–96.

John, S. (2018): "Epistemic Trust and the Ethics of Science Communication: Against Transparency, Openness, Sincerity and Honesty", *Social Epistemology* 32, 75–87.

Kant, I. (1785): *Groundwork for the Metaphysics of Morals* (tr. T. K. Abbott, ed. L. Denis), Broadview Press 2005.

Kincaid. H., Dupré, J., and Wylie, A. (eds, 2007): *Value-free Science?*, Oxford University Press.

Lacey, H. (2013): "Rehabilitating Neutrality", *Philosophical Studies* 163, 77–83.

Le Bihan, S. (2024): "How Not to Secure Public Trust in Science: Representative Values vs. Polarization and Marginalization", appears in *Philosophy of Science*.

MacIntyre, A. (1981): *After Virtue. A Study in Moral Theory*, Dickworth.

Mill, J. S. (1863): *Utilitarianism*, Parker, Son, and Bourn.

Moore, G. E. (1903): *Principia Ethica*, Cambridge Univ. Press.

NAS (National Academies of Sciences) (2018): "Open Science by Design: Realizing a Vision for 21st Century Research", *The National Academies Press*.

Nozick, R. (1974): *Anarchy, State and Utopia*, Basil Blackwell, Oxford.

Parker, W. (2024): "The Epistemic Projection Approach to Values in Science", *Philosophy of Science* 91, 18–36.

Pigden, C. R. (1989): "Logic and the Autonomy of Ethic", *Australasian Journal of Philosophy* 67, 127–151.

Prior, A. (1960): "The Autonomy of Ethics", *Australasian Journal of Philosophy* 38, 199–206.

Putnam, H. (2002): *The Collapse of the Fact/Value Dichotomy*, Harvard University Press.

Rawls, J. (1971): *A Theory of Justice*, Cambridge Univ. Press, Cambridge.

Reichenbach, H. (1956): *The Direction of Time*, Univ. of California Press, Berkeley.

Rudner, R. (1953): "The Scientist qua Scientist Makes Value Statements", *Philosophy of Science* 20, 1–6.

Russell, G. (2022): "How to Prove Hume's Law", *Journal of Philosophical Logic* 51, 603–632.

Scheler. M. (1913/16): *Der Formalismus in der Ethik und die materiale Wertethik* (hg. C. Bermes), Felix Meiner Verlag 2014.

Schmidt, P. F. (1971): "Ethische Normen in der wissenschaftlichen Methode", in: Albert/Topitsch (eds., 1971), 353–364.

Schooler, J. (2011): "Unpublished Results Hide the Decline Effect", *Nature* 470, 437.

Schroeder, S. (2021): "Democratic Values: A Better Foundation for Public Trust in Science", *The British Journal for the Philosophy of Science* 72, 545–562.

Schurz, G. (1997): *The Is-Ought Problem. An Investigation in Philosophical Logic*, Kluwer 1997.

Schurz, G. (2010): "Non-Trivial Versions of Hume's Is-Ought Thesis and Their Presuppositions", in: Pigden, C. R. (ed., 2010): *Hume on 'Is', and 'Ought'*, Palgrave Macmillan, 198–216.

Schurz, G. (2013): "Wertneutralität und hypothetische Werturteile in den Wissenschaften", in: Schurz, G. and Carrier, M. (Eds., 2013), *Werte in den Wissenschaften*, Suhrkamp, 305–334.

Schurz, G. (2014): *Philosophy of Science: A Unified Approach*, Routledge, New York.

Schurz, G., and Gebharter, A. (2016): "Causality as a Theoretical Concept", *Synthese* 193 (4), 1071–1103.

Searle, J. (1964): "How to Derive 'Ought' from 'Is'", *Philosophical Review* 73, 43–58.

Spirtes, P., Glymour, C., and Scheines, R. (2000): *Causation, Prediction, and Search*, MIT Press.

Steele, K. (2012): "The Scientist Qua Policy Advisor Makes Value Judgements", *Philosophy of Science* 79, 893–904.

Strauss, L. (1971): "Die Unterscheidung von Tatsachen und Werten", in: Albert (Topitsch (eds., 1971), 73–91.

Vom Saal, F. S., and Hughes, C. (2005): "An Extensive New Literature Concerning Lowdose Effects of Bisphenol A Shows the Need for a New Risk Assessment", *Environmental Health Perspectives* 113, 926–933.

Weber, M. (1917): "The Meaning of 'Ethical Neutraliy' in Sociology and Economics" (Engl. trans.), in: *Max Weber in the Methodology of the Social Sciences* (ed. E. A. Shils and H. A. Finch), Free Press 1949, 1–49.

Wiesing, U. et al. (2021): "Wissenschaftliche (Politik-)Beratung in Zeiten von Corona: Die Stellungnahmen der Leopoldina zur Covid-19-Pandemie", in: *Ethik und Gesellschaft* 1 (doi: 10.18156/eug-1-2021-858).

Wilholt, T. (2009): "Bias and Values in Scientific Research", *Studies in History and Philosophy of Science* 40, 92–101.

www.ingramcontent.com/pod-product-compliance
Lightning Source LLC
Chambersburg PA
CBHW060520090426
42735CB00011B/2310